Autonomous Learning Systems

Autonomous Learning Systems

From Data Streams to Knowledge in Real-time

Plamen Angelov

Lancaster University, UK

This edition first published 2013

Registered office
John Wiley & Sons Ltd, The Atrium, Southern Gate, Chichester, West Sussex, PO19 8SQ, United Kingdom

For details of our global editorial offices, for customer services and for information about how to apply for permission to reuse the copyright material in this book please see our website at www.wiley.com.

Library of Congress Cataloging-in-Publication Data:

Angelov, Plamen P.
Autonomous learning systems : from data streams to knowledge in real-time / Plamen P. Angelov.
pages cm
Includes bibliographical references and index.
ISBN 978-1-119-95152-0 (cloth)
1. Self-organizing systems. 2. Machine learning. I. Title.
Q325.A54 2013
006.3′1–dc23

2012025907

A catalogue record for this book is available from the British Library.

ISBN: 978-1-119-95152-0

Set in 10/12.5pt Palatino by Aptara Inc., New Delhi, India

Printed and bound in Malaysia by Vivar Printing Sdn Bhd

Contents

PART III APPLICATIONS OF ALS

Forewords

Adrian Stoica

Efficient and robust performance in imperfectly known, nonstationary, environments – and this characterizes the vast majority of real-world applications – requires systems that can improve themselves, transcending their initial design, continuously optimizing their parameters, models, and methods. These improvements come predominantly from learning – about the environment, about the ageing self, about the interactions with, and within, the environment, and from the ability to put this learning to use. Batch learning – or at least repeated updating learning from most recent batches, is sufficient only for a limited number of applications. For other applications learning needs to be incremental, to sample level, a learn-or-perish, or at least learn-or-pay (a hefty price) situation. In particular, real-time learning is most critical for bots, virtual or real, agents of the cyberphysical systems that need the agility to swiftly react to virus attacks, or physical robots exposed to hazards while performing search and rescue in disaster areas, or dealing with what is, for now, a largely unpredictable partner: the human. The fast advancement in autonomous systems makes the subject of real-time autonomous learning critically important, and yet the literature addressing this topic is extremely scarce.

Dr Angelov's pioneering book addresses this problem at its core, focusing on real-time, online learning from streaming data on a sample-by-sample basis. It offers a basic framework for understanding and for designing such systems. It importantly contributes to a more powerful learning, not only of the parameters but of a better structure as well. Conventional approaches are characterized by the fact that the system structure (model) is determined in the beginning, by human designers of the system, and only parameters are learned from the interaction. The entire model identification–learning process can, however, be posed as an optimization problem, as the author points out, and this includes automatically determining the optimal structure in conjunction with the optimal parameters for it. This is done automatically in the methods described in the book, and constitutes a significant and valuable contribution. The system is continuously evolving, not in the evolutionary (genetic) sense of improvement over generations, but continuously perfecting its development.

A valuable contribution of the book is that it offers a high-level, holistic perspective of the field, which helps both students and expert practitioners better comprehend the interplay of various disciplines involved in learning autonomous systems, as diverse as adaptive control and evolutionary algorithms, offering analogies between different disciplines and referring to the equivalency of the concepts characterized by different terminology in different disciplines. It is not meant to be a comprehensive reference of concepts and methods in the field, the author instead paints the landscape with selected brush strokes that allows the viewer to see the forest without getting lost in seeing the trees. It is a work that charters a new field, innovates in methods to advance into it, and outlines new challenges to be addressed by future explorers.

The selected concepts and methods, a good number of which come from the author's own prior work, are used in the second part of the book to illustrate the implementation of learning in autonomous systems. Concepts such as that of evolving clusters, 'age' of an (evolving) local submodel, and methods such as recursive density estimation (RDE) introduced by the author, showing significant improvement over the state-of-the art, are important additions to the arsenal of tools for real-time learning. In particular, I believe that adaptive, self-learning (evolving) classifiers will play a fundamental role in future autonomous learning systems.

The book's last part is a review of three applications: autonomous learning sensors for chemical and petrochemical applications, autonomous learning in mobile robots, and autonomous novelty detection and object tracking in video systems. These diverse domains illustrate the general applicability of the set of methodology presented in the book and focused on the main theme of this work: the real-time autonomous online learning from data streams.

The field of autonomous learning systems is destined to play an increasingly important role in most systems that will surround us in cyberphysical space. Converting information in data streams, larger and larger, to actionable knowledge, in real time: this is the great challenge ahead, and this book is an important step towards addressing it.

Adrian Stoica
Jet Propulsion Laboratory
California Institute of Technology
June 2012

Vladik Kreinovich

In many practical situations, we have experts who are skilled in doing certain tasks: expert medical doctors are skilled in diagnosing and curing diseases, professional drivers are skilled in driving cars – in particular, driving them in difficult traffic and/or weather conditions, etc. It is desirable to incorporate the knowledge of these top experts in an automatic system that would help other users perform the corresponding tasks – and, ideally, perform these tasks automatically.

Experts are usually willing to share their knowledge, but the difficulty is that in many situations, experts describe their knowledge by using imprecise ("fuzzy") words from natural language, like "small". For example, an expert driver rarely describes his or her experience in precise terms like "if the car in front slows down by 10 km/h and it is at a distance of 10 meters, you should press the brake for 0.6 seconds with a force of 2.7 Newtons"; most probably, the rule described by an expert driver is "if the car in front of you is close, and it suddenly slows down some, then you should brake right away". In this rule, "close" and "some" are imprecise terms: while everyone would agree that, say 100 meters is not close while 5 meters is close, there will not be a precise threshold so that before this threshold the distance is close, and a 1 cm larger distance is not close.

To describe such imprecise (fuzzy) knowledge in computer-understandable precise terms, Professor Lotfi A. Zadeh invented, in the 1960s, a new approach called *fuzzy logic*. Zadeh's ideas led to *revolutionary* changes in many control situations: from the first successful control applications in the 1970s through the fuzzy control boom in the 1990s – when fuzzy-controlled washing machines, camcorders, elevators, trains were heavily promoted and advertised – to the current ubiquity of fuzzy controllers. Just like nowadays computers are ubiquitous – companies no longer brag about computer chips in their cars, since all the cars have such chips – similarly, fuzzy control is ubiquitous: for example, in many cars, automatic transmission systems use fuzzy control.

The existing fuzzy controllers are very successful, but they have a serious limitation: they do not learn. Once the original expert rules are implemented, these same rules are used again and again, even when it becomes clear that the rules need to be updated. We still need an expert to update these rules.

There are, of course, numerous intelligent systems that *can* learn, such as artificial neural networks, but from the viewpoint of the user, these systems are "black boxes": we may trust them, but we cannot easily understand the recommendations. In contrast, fuzzy rules, by definition, are formulated in terms of understandable rules. If we could make fuzzy systems themselves learn, make them automatically update the rules – this would combine the clarity of fuzzy rules with the autonomous learning ability of neural networks. This would make learning fuzzy controllers even more efficient – and therefore, even more widely used. This would lead to a *second revolution* in intelligent control.

And this revolution is starting. This book, by Dr. Plamen Angelov, one of the world's leading specialists in learning fuzzy systems, is the first book that summarizes the current techniques and successes of autonomously learning fuzzy (and other) systems – techniques mostly developed by Dr. Angelov himself, often in collaboration with other renowned fuzzy researchers (like Dr. Ronald Yager). Some of these techniques have previously appeared in technical journals and proceedings of international conferences, some appear here for the first time.

Ideas are many, it is difficult to describe them all in a short preface, so let us just give a few examples. The first example is an interesting AnYa algorithm invented by *An*gelov and *Ya*ger (Anya is also a Russian short form of Anna (Anne)). In fuzzy logic, each fuzzy term like "small" is described by a *membership function*, i.e. a function that assigns, to each possible value x, the degree $\mu(x)$ from the interval [0, 1] to which this value is small. The value $\mu(x) = 1$ means that x is absolutely small, every expert would agree to this; $\mu(x) = 0$ means that x is definitely not small, while values between 0 and 1 represent the expert's uncertainty.

In the traditional fuzzy control algorithms, we select a finite-parametric family of membership functions – e.g., functions that are of triangular, trapezoid, or Gaussian shapes – and adjust parameters of these functions based on the expert opinions. As a result, sometimes, the resulting membership functions provide a rather crude and not very accurate description of the expert knowledge. To improve the situation, AnYa does not limit the shape of the membership function. Instead, it uses all the value $x_1, \ldots, x_n$, that the expert believes to be satisfying the property (like "small"), and defines the desired membership function by formalizing the statement "x is close to x_1 or x is close to x_2, ...". Now, all we need to do is describe what experts mean by "close" (and by "or"), and we will no only have a well-shaped membership function, we will also have a way to update its shape when new observations appear.

A similar idea can be implemented in probabilistic terms, when we use probability density functions (pdf) $\rho(x)$ instead of membership functions, but the authors show a clear computational advantage of their fuzzy approach: A pdf is normalized by the condition that the total probability is 1: $\int \rho(x)dx = 1$, so we need to go through a computationally intensive process of renormalize all its values every time we update one value of $\rho(x)$. In contrast, a membership function is usually normalized by the condition that $\max_x \mu(x) = 1$. Thus, if we change a value of the membership function, we only need to renormalizing other values when the changed value is $\mu(x) = 1$ – and this happens rarely.

Similar ideas are used to automatically decide how to adjust the rule's conclusions, when to subdivide the original rule into two subrules – that would provide a more subtle description of actions, when to dismiss the old data that is no longer representative of the system's inputs, etc.

Researchers and practitioners who have been using fuzzy techniques will definitely benefit from learning how to make fuzzy systems learn automatically (pun intended :-). But this book is not only for them. Readers who are not familiar with the current fuzzy techniques will also greatly benefit: the book starts with a nice introduction that explains, in popular terms, what is fuzzy, and why and how we can use fuzzy techniques. (Some math is needed – but math taught to engineers is quite enough.)

This book is not just for the academics, practitioners will surely benefit. In the last part of the book, numerous applications are described in detail, providing the reader with an understanding of how these new methods can be used in practical situations. It may be a good idea to glance through these exciting applications first, this will give the readers an excellent motivation to grind through all the formulas and algorithms in the main part of the book.

Applications include learning sensors for chemical and petrochemical industries – industries where the chemical contents of the input (such as oil) changes all the time, and intelligent adjustments need to be constantly made. Another successful application example is mobile robotics, where the robot's ability to learn how to navigate in a new environment – and learn fast – is often crucial for the robot's mission. The new methods have also been applied to novelty detection and object tracking in video streams, to wireless sensor networks, and to many other challenging application areas.

The second revolution – of making intelligent control systems fast learners – has started. Its preliminary results are already exciting. This book will definitely help promote the ideas of this second revolution – and thus, further improve its methods and use these improved methods to solve numerous challenging problems of today.

Vladik Kreinovich
President
North American Fuzzy Information Processing Society (NAFIPS), El Paso, Texas
August 2012

Arthur Kordon

One of the most significant changes during the twenty-first century is the fast dynamics in almost all components of life. Economic, social, and financial systems, to name a few, are moving more of their activities to a real-time mode of operation. Extracting knowledge from data streams becomes as important as was information retrieval from data bases several decades ago. The need for fast adaptation to unknown conditions, due to the new complex nature of the global economy, is another big challenge in operating the systems in the twenty-first century.

Unfortunately, the existing classical modeling techniques, based on first principles, statistics, system identification, etc., cannot deliver satisfying solutions adequate to the new fast dynamics. Adaptive systems are limited to models with a fixed structure and linear relationships. Some recent computational intelligence methods with nonlinear and adaptive capabilities, such as evolutionary computation and swarm intelligence, are too slow for real-time operation.

A potential solution to the new needs is the fast-growing research area of evolving intelligent systems. They offer a system that simultaneously learns its structure and calculates its parameters "on the fly" from data streams. An advantage of this approach is the simplicity and very low memory requirement of the used algorithms, which makes them appropriate for real time. In addition, the algorithms are universal (i.e. can be applied in various areas with no or minor changes), with minimal number of tuning parameters, and in the case of evolving fuzzy systems – the suggested models are interpretable by the users. Autonomous learning systems (ALS) is the ultimate solution of an evolved intelligent system that integrates the broadest possible range of algorithms and requires minimal human intervention. The research area of ALS has grown significantly in terms of publications, conference presence, and funding support. An impressive feature of this emerging technology is its fast applicability in different areas, such as inferential sensors, mobile robotics, video streams processing, defence applications, etc.

Unfortunately, the literature for ALS is mostly available in journal or conference papers. Some recently published books on evolving intelligent systems and evolving fuzzy systems give a generic overview and some guidelines on the different components of the technology. However, researchers and practitioners need a book that describes in sufficient details the foundation of ALS, offers software for investigating the key algorithms in a popular environment, such as Matlab, and gives appropriate application examples. *Autonomous Learning Systems* is the first book on the market that fills this need.

Purpose of the Book

The purpose of the book is to give the reader a comprehensive view of the current state of the art of ALS. The key topics of the book are:

1. *What are the fundamentals of ALS?* The first main topic of the book focuses on the ambitious task of describing the diverse research foundation of ALS. The key methods, such as probability theory, machine learning and pattern recognition, clustering, and fuzzy system theory are presented at a level of detail sufficient for understanding the ALS mechanisms.
2. *How to develop an ALS?* The second key topic of the book is the description of the methodology of ALS. Its main focus is on presenting the key steps in developing of an ALS from streaming data, such as dynamic data space partitioning, normalization and standardisation, autonomous monitoring of the structure quality, and autonomous learning parameters of the local submodels.
3. *How and where to apply ALS in practice?* The third key topic of the book covers the implementation issues and application areas of ALS. It includes an overview of the potential application areas, such as autonomous predictors, estimators, filters, and inferential sensors; autonomous learning classifiers; autonomous learning controllers and collaborative autonomous learning systems. Of special interest are the results from several real-world applications of ALS, such as inferential sensors in the chemical and petrochemical industries, ALS in mobile robotics, and autonomous novelty detection and object tracking in video streams.

Who this Book is for?

Due to the wide potential application areas, the targeted audience is much broader than the traditional scientific communities in computer science, data mining, and engineering. The readers who can benefit from this book are described below:

- *Academics* – This group includes a large class of academics in the fields of computer science, data mining, and engineering who are not familiar with the research and technical details of this new field. They will benefit from the book by using it as an introduction to the field, exploring the described algorithms, and understanding its broad application potential.
- *Students* – Undergraduate and graduate students can benefit from the book by understanding this new field. The book could be a basis for a graduate course on this topic.
- *Industrial researchers* – This group includes scientists in industrial labs who create new solutions for the busines. They will benefit from the book by understanding the value of this new emerging technology in delivering novel solutions in the area of real-time modelling based on data streams.
- *Governmental agencies* – ALS have almost unlimited potential in various military applications and space exploration projects. This book can be used by the governmental decision makers as an introduction to the technology and even may open new application areas.
- *Software vendors* – This group includes vendors of process monitoring and control systems, data-mining software, robotic systems, etc. They will benefit from the book by understanding a new emerging technology, exploring the described algorithms, and defining new application areas, related to data streams.

Features of the Book

The key features that differentiate this book from other titles on learning and evolving systems are:

1. *A systematic description of autonomous learning systems* – One of the most valuable features of the book is the systematic and comprehensive way of presenting related methods. It gives the reader a solid basis for understanding the research foundation of ALS, which is of critical importance for introducing this emerging technology to a broad audience.
2. *A detailed methodology for development of autonomous learning systems* – Another important topic of the book is the step-by-step description of the key algorithms, used in the proposed technology. This allows the reader to easily implement and explore the large potential of ALS on a simulation level.
3. *A broad range of autonomous learning systems applications* – The third key feature of the book is the impressive list of described real-world applications across several application areas. It illustrates one of the unique advantages of ALS – the fast transition from theory to practice.

Arthur Kordon
Lake Jackson, Texas
March, 2012

Lawrence O. Hall

Today's data-driven world has many large streams of data, for example a day's worth of images posted to the internet. Imagine the task of finding all instances, perturbed or not, of an image. It would require learning and adaptation as the way images are placed and modified evolve. This book provides an informative snapshot of how to build autonomous learning systems for data streams. It covers the math you need, probability, normalization, fuzzy systems, the basics of machine learning and pattern recognition, clustering, feature selection and more. It shows how to do learning in an autonomous setting and how to tune out noise/outliers.

Sensor learning, autonomous classifier learning, collaborative learning, learning controls systems in an autonomous way are all covered in an interesting way. There are case studies given to show how it all fits together. These include object tracking in videos and autonomous learning robots. It is clear that we need adaptable robots capable of modifying their behaviour. This is an ambitious, early book that covers robot learning and the basics for much more autonomous learning. It is well worth perusing for those interested in this broad subject.

Lawrence O. Hall
Dept. of Computer Science and Engineering, University of South Florida, USA
May 16, 2012

Preface

This book comes as a result of focused research and studies for over a decade in the emerging area that is on the crossroads of a number of well-known and well-established disciplines, such as (Figure 1):

- machine learning (ML);
- system engineering (specifically, system identification), SI;
- data mining, DM;
- statistical analysis, SA;
- pattern recognition including clustering, classification, PR;
- fuzzy logic and fuzzy systems, including neurofuzzy systems, FL;
- and so on.

On the one hand, there is a very strong trend of innovation of all of the above well-established branches of research that is linked to their *online* and *real-time* application; their adaptability, flexibility and so on (Liu and Meng, 2004; Pang, Ozawa and Kasabov, 2005; Leng, McGuinty and Prasad, 2005). On the other hand, a very strong driver for the emergence of *autonomous learning systems* (ALS) is industry, especially defence and security, but also aerospace and advanced process industries,

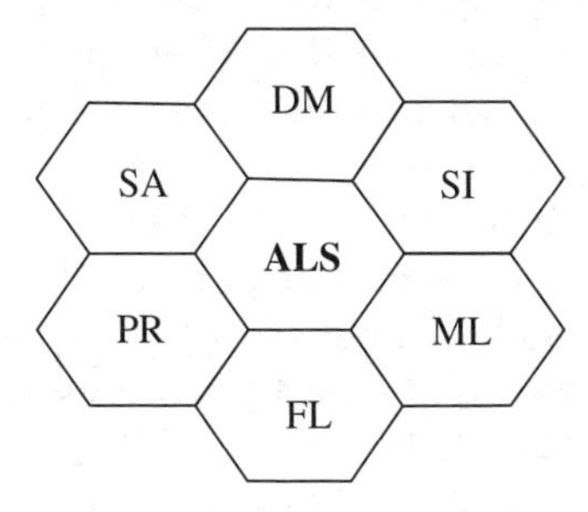

Figure 1 Autonomous learning systems theory is build upon other well-established areas of research (the list is, of course, not exhaustive)

the Internet, eHealth (assisted living), intelligent transport, and so on. The demand in defence was underpinned recently by a range of multimillion research and development projects funded by DARPA, USA (notably, two Grand Challenge competitions (Buehler, Iagnemma and Singh, 2010)); by MoD and BIS, UK (Defence Technology Centre on Systems Engineering and Autonomous Systems; ASTRAEA and GAMMA multimillion programmes, in which the author has played the research provider role, being the technical lead for several projects) and similar programmes in other European countries (France, Sweden, Spain, Czech Republic, Russia), and Israel. Major global companies have established their own programmes, such as IBM's autonomous computing initiative (IBM, 2009) and BT's intelligent network of autonomous elements (Detyenecki and Tateson, 2005). The International Neural Network Society (INNS) has established in 2011 a section on Autonomous Machine Learning of which the author is a founding member, together with scientists such as Bernard Widrow – the father of the famous least mean squares (LMS) algorithm.

This book attempts to address these challenges with a systematic approach that can be seen as laying the foundations of what can become a fast-growing area of research that can underpin a range of technological applications so needed by industry and society. The author does not claim that this represents a finished and monolithic theory; this is rather a catalyser for future developments, an inoculum, a vector pointing the direction rather than a full solution of the problems.

An important aim of preparing this book was also to make it a one-stop shop for students, researchers, practicing engineers, computer specialists, defence and industry experts and so on that starts with the motivation, presents the concept of the approach, describes details of the theoretical methodology based on a rigorous mathematical foundation, presents a wide range of applications, and more importantly, provides illustrations and algorithms that can be used for further research. The software (subject to a license) can be downloaded from the author's web site: http://www.lancs.ac.uk/staff/angelov/Downloads.htm. It is covered by USA patent # 2010-0036780, granted 21 August 2012 (priority date 1 Nov. 2006) and two pending patent applications and distributed by the spin-out company of Lancaster University called EntelSenSys Ltd. (www.entelsensys.com). From the same web site there will also be available for the readers for this book a set of lecture notes that will be a useful tool for delivering specialised short courses or an advanced Master level module as a part of various related programmes that cover the topics of machine learning, pattern recognition, control systems, computational intelligence, data mining, systems engineering, and so on.

The book was initially planned at the end of 2006 during the very successful IEEE Symposium on Evolving Fuzzy Systems held in Ambleside in the Lake District, UK but the turn of events (as usually happens) postponed its appearance by more than five years, which gave an opportunity for the concepts to mature and evolve further and new results and applications to be added.

It would not have become a reality without the support of the colleagues and collaborators, students, associates and visitors of the author. In the hope not to miss someone this includes Prof. Ronald Yager (Iona College, NY, USA), Dr. Dimitar Filev

(Ford, MI, USA), Prof. Nikola Kasabov (Auckland University, New Zealand), Prof. Fernando Gomide (University of Campinas, Brazil), Dr. Xiaowei Zhou (HW Communications, UK), Dr. Jose Antonio Iglesias (University Carlos III, Madrid, Spain), Dr. Jose Macias Hernandez (CEPSA, Tenerife, Spain), Dr. Arthur Kordon (The Dow Chemical, TX, USA), Dr. Edwin Lughofer (Johan Kepler University, Linz, Austria), Prof. Igor Skrjanc (University of Ljubljana, Slovenia), Prof. Frank Klawonn (Ostfalia University of Applied Sciences, Germany), Mr. Jose Victor Ramos (University of Coimbra, Portugal), Dr. Ana Cara Belen (University of Granada), Mr. Javier Andreu (Lancaster University), Mr. Pouria Sadeghi-Tehran (Lancaster University), Mr. Denis Kolev (Rinicom Ltd.), Mrs. Rashmi Dutta Baruah (Lancaster University), Mr. Ramin Ramezani (Imperial College, London), Mr. Julio Trevisan (Lancaster University), and many more.

The feedback on the manuscript by Professor Vladik Kreinovich (University of Texas, USA) who is also President of the North American Fuzzy Information Processing Society (NAFIPS); Dr. Adrian Stoica, Senior Research Scientist at the Autonomous Systems Division, NASA Jet Propulsion Laboratory, Pasadena, CA, USA, Dr. Arthur Kordon, Team Leader at Dow Chemical, TX, USA; as well as from Dr. Larry Hall, Distinguished Professor and Chair at the Department of Computer Science and Engineering, University of South Florida, USA was also instrumental to improve and smooth out the presentation and remove some omissions.

Plamen Angelov
November 2009–May 2012
Lancaster, UK

About the Author

The author, *Dr Plamen Angelov*, is a Reader in Computational Intelligence and coordinator of the Intelligent Systems Research Area at Infolab21, Lancaster University, UK. He is a Senior Member of the IEEE and of INNS (International Neural Networks Society) and Chair of the Technical Committee on Evolving Intelligent Systems, Systems, Man and Cybernetics Society, IEEE. He is also a member of the UK Autonomous Systems National Technical Committee and a founding member of the Centre of Excellence in CyberSecurity officially recognised by UK GCHQ for the period 2012–2017.

He has authored or coauthored over 160 peer-reviewed publications in leading journals (50+) peer-reviewed conference proceedings, three patents, two research monographs, half a dozen edited books, and has an active research portfolio in the area of computational intelligence and autonomous system modelling, identification, and machine learning. He has internationally recognised pioneering results into online and evolving methodologies and algorithms for knowledge extraction in the form of human-intelligible fuzzy rule-based systems and autonomous machine learning.

Dr. Angelov is a very active researcher leading numerous projects (over fifteen in the last five to six years) funded by UK and EU research councils, industry, HM Government, including UK Ministry of Defence (total funding of the order of tens of millions pounds with well over £1M for his group alone). His research contributes to the competitiveness of the industry, defence and quality of life and was recognised by 'The Engineer Innovation and Technology 2008 Award in two categories: i) Aerospace and Defence and ii) The Special Award.

Dr. Angelov is also the founding Editor-in-Chief of Springer's journal on *Evolving Systems* and Associate Editor of the leading international scientific journals in this area, including *IEEE Transactions on Systems, Man and Cybernetics*, *IEEE Transactions on Fuzzy Systems*, Elsevier's *Fuzzy Sets and Systems*, *Soft Computing*, *Journal on Automation, Mobile Robotics and Intelligent Systems* Journal on Advances in Aircraft and Spacecraft Science and several others. He also chairs annual conferences organised by IEEE on Evolving and Adaptive Intelligent Systems, will be General co-Chair of the prime conferences on neural networks (IJCNN-2013, Dallas, Texas, August, Texas and fuzzy systems, FUZZ-IEEE-2014, June 2014, Beijing, China and on Cybernetics, CYBCO-2013, Lausanne, Switzerland), acted as Visiting Professor (in Brazil, Germany, Spain) regularly gives invited and plenary talks at leading conferences, universities and companies More information can be found at his web site www.lancs.ac.uk/staff/angelov.

The *evolving* face of the author (who is, of course an autonomous learning and evolving system himself) can be seen below:

1970 1982 1999 recently

1

Introduction

The main differentiator of the new generation of autonomous systems that is emerging in the twenty-first century is the *adaptivity* of their intelligence. They are not simply automatic (usually remote) control devices, not only adaptive control systems in the narrow sense of systems with *tunable* parameters as in the last decades of the past century, but they are rather systems with a certain level of evolving intelligence. While conventional adaptive techniques (Astroem and Wittenmark, 1989) are suitable to represent objects with slowly changing parameters, they can hardly handle complex (usually, nonlinear, nonstationary) systems with multiple operating modes or abruptly changing characteristics since it takes a long time after every drastic change in the system to update model parameters. The *evolving* systems paradigm (Angelov, 2002) is based on the concept of evolving (expanding or shrinking) system *structure* that is capable of adapting to the changes in the environment and internal changes of the system itself that cannot solely be represented by parameter tuning/ adjustment.

Evolving intelligent systems (eIS) the concept of which was pioneered recently (Angelov, 2002; Kasabov, 2002; Angelov and Kasabov, 2005; Kasabov and Filev, 2006, Jager, 2006), develop their structure, their functionality, and their internal knowledge representation through *autonomous learning* from data *streams* generated by the (possibly unknown) environment and from the system self-monitoring. They often (but not necessarily) use as a framework of implementation fuzzy rule-based (FRB) and neurofuzzy (NF) or neural-network (NN) based systems and machine learning as a tool for training. Alternative frameworks (such as conventional multimodel systems, decision trees, probabilistic, e.g. Markov, mixture Gaussian models, etc.) can also be explored as viable frameworks of eIS and *autonomous learning* systems.

It should be noted that the physical embodiments of such systems can range from micro-systems-on chip (Everett and Angelov, 2005), motes of a wireless sensor network (Andreu, Angelov and Dutta Baruah, 2011), mobile robots (Zhou and

Autonomous Learning Systems: From Data Streams to Knowledge in Real-time, First Edition. Plamen Angelov.

Angelov, 2007; Liu and Meng, 2004; Kanakakis, Valavanis and Tsourveloudis, 2004) to unmanned airborne vehicles (Valavanis, 2006) and computer-controlled industrial processes (Filev, Larson and Ma, 2000; Macias-Hernandez *et al.*, 2007).

The potential of these systems for industry was acknowledged by leading researchers with a solid industrial background such as Dr. A. Kordon, R&D Leader, Dow Chemical, TX, USA who said in 2006 "*Evolving Intelligent Systems have a high potential for implementation in industry*" (http://news.lancs.ac.uk/Web/News/Pages/930389757F5B0BF4802571FB003CB1A2.aspx); Dr. D. Filev, Senior technical Staff at Ford R&D, Dearborn, MI, USA who also in 2006 said "*Embedded soft computing applications are the natural implementation area of evolving systems as one of the main tools for design of real time intelligent systems*" (same web reference as above).

The problem of automatic design of computationally intelligent systems for modelling, classification, time-series prediction, regression, clustering from data has been successfully addressed during the previous century by a range of techniques such as by gradient-based techniques (as in the neurofuzzy approach ANFIS (Jang, 1993)), by genetic/evolutionary algorithms (Fogarty and Munro, 1996; Angelov and Buswell, 2003), by using partitioning by clustering (Babuska, 1998), learning by least squares (LS) techniques and so on. But, these approaches were assuming all data to be known in advance (a *batch* or *offline* mode of learning) and were not directly applicable to data *streams*.

At the same time, the twenty-first century is confronting us with a range of new challenges that require completely new approaches. As John Naisbitt famously said "today we are drowning in information but starved for knowledge" (Naisbitt, 1988). We are in the midst of an information revolution witnessing an exponential growth of the quantity and the rate of appearance of new information by; Internet users, consumers, finance industry, sensors in advanced industrial processes, autonomous systems, space and aircraft, and so on.

It is reported that every year more than 1 Exabyte ($=10^{18}$ bytes) of data are generated worldwide, most of it in digital form (http://news.bbc.co.uk/2/hi/technology/4079417.stm). Toshiba recently coined the phrase '*digital obesity*' to illustrate the ever-growing amount of data that are generated, transmitted and consumed by the users today. In this ocean of data the useful information and knowledge very often is difficult to extract in a clear and comprehensive, human-intelligible form. The availability of convenient-to-use and efficient methods, algorithms, techniques, and tools that can assist in extracting knowledge from the data (Martin, 2005) is a pressing demand at individual and corporate level, especially if this can be done online, in real time.

The new challenges that cannot be successfully addressed by the existing techniques, especially in their complexity and interconnection, can be summarised as follows:

i. to cope with huge amounts of data;
ii. to process streaming data online and in real time;
iii. to adapt to the changing environment and data pattern autonomously;

iv. to be computationally efficient (that means, to use recursive, one-pass, noniterative approaches);
v. to preserve the interpretability and transparency in a dynamic sense.

To address these new challenges efficient approaches are needed that deal with data streams (Domingos and Hulten, 2001), not just with batch sets of data (Fayyad *et al.*, 1996), detect, react and take advantage of concept *shift* and *drift* in the data streams (Lughofer and Angelov, 2011). Efficient collaborative and interactive schemes are also needed for a range of applications in process industries (for self-calibrating, self-maintaining intelligent sensors of the new generation), in autonomous systems and robotics (for systems that have self-awareness, replanning and knowledge summarisation capabilities), in multimedia and biomedical applications, to name a few.

Autonomous learning (AL) is understood in this book in the context of both system structure and system parameters. This means that the overall process of design, development, redesign/update, adaptation, use and reuse of such systems is autonomous, including the stages of the system design that traditionally assume heavy human involvement and are normally done *offline* (the system being designed not in *real time* in which the process that is using this system runs in). Therefore, our understanding of AL and our concept of eIS is intricately related to the concepts of *online* and *real-time* system structure and parameter design and exploitation and to data *streams* rather than to data *sets*. This is the main differentiator in comparison with the traditional disciplines.

1.1 Autonomous Systems

Autonomous systems are often seen as the physical embodiments of machine intelligence. The concept of autonomous systems (AS) is not new and is closely related to AI and cybernetics, but became more popular during the last decade or so mainly due to the interest (and funding) from the defence and aerospace industries. AS are significantly different from simple automatic control systems, ACS (Astroem and Wittenmark, 1989). In fact, each AS has at its lower level (Layer 1) an ACS, usually, for motion control, for control of the sensors and actuators, and so on. An AS, however, has also important upper layers in its architecture (Figure 1.1) that concern perceptions-behaviours (layer 2 that also corresponds to structure identification in AL systems) and the representation of the environment (usually in a form of rule base, states, or a map, but not necessarily limiting to these forms of representation) in the model and self-monitoring functions (layer 3 that is linked to the prediction).

AS can be seen as a fusion of computationally enabled sensor platforms (machines/devices) that possess the algorithms (respectively, the software) needed to empower the systems with evolving intelligence that is manifested through interaction with the outside environment and self-monitoring.

Examples include, but are not limited to unmanned airborne systems, UAS, unmanned ground-based vehicles, UGVs (Figure 1.3), and so on.

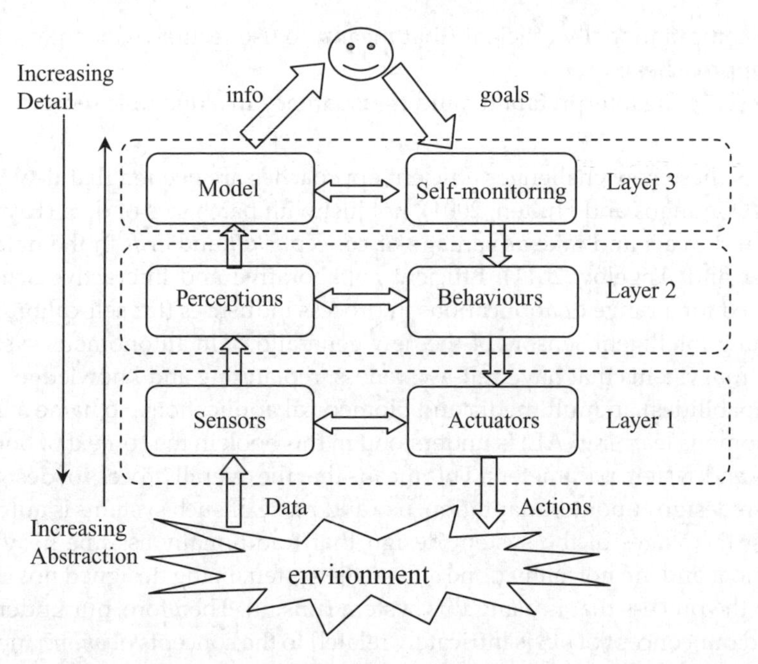

Figure 1.1 A three-layer structure of an autonomous system. (layer 1 – low-level direct control, including teleoperation; layer 2 – a more abstract, behavioural autonomy, specific tasks; layer 3, often called deliberate autonomy – the upper abstract layer of modelling the environment and self-monitoring)

1.2 The Role of Machine Learning in Autonomous Systems

The core functionality of an AS depends on the ability to be aware of the environment (through data streams generated by the sensors) and to make decisions accordingly. Obviously, such decisions cannot be made on the basis of a preprogrammed logic because this will assume a full knowledge of all the environments in which the system will operate and will not be flexible enough. Therefore, core elements of any AS are self-monitoring and self-adaptation. Autonomous learning and extracting new knowledge as well as updating the existing knowledge base are vitally important for such type of systems.

The dependence between autonomy and learning is a two-way process – on the one hand, autonomous systems require learning in order to be aware of, explore, and adapt to the dynamic environment; on the other hand, learning algorithms require autonomy to make them independent of human involvement. The lack or low level of autonomy in most of the currently existing algorithms leads to the need to develop new generations of AL systems (ALS) to play an important role in the design and maintenance of autonomous systems (e.g. UAV, UGV, intelligent/soft sensors, etc.). A system (however well may it have been designed) that is not empowered by an autonomous machine learning capability will fail helplessly in a situation that was

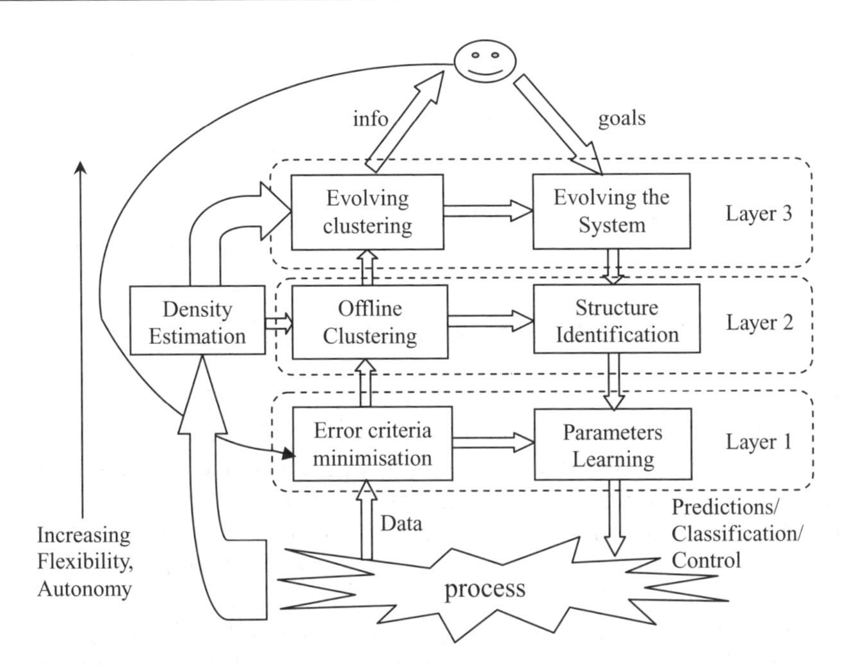

Figure 1.2 A graphical representation of an autonomous machine learning system (layer 1 – 'traditional' (parameters only learning) approach; layer 2 – offline learning of system structure using clustering and data density (it should be noted that other methods instead of density-based clustering can be used at this layer); layer 3 represents the evolving system structure – the upper layer of autonomous learning that often also includes self-monitoring (not represented for simplicity)

not predicted at the design stage or a situation that is described by parameters widely out of the range of parameters considered during the design stage.

A system that has learning capabilities and an evolving model of the real world will try to adapt and create new rules, to drop rules that are outdated and irrelevant to the new situation and will at least have a higher chance to succeed. In reality, most of the complex environments are unpredictable, nonlinear and nonstationary. An

Figure 1.3 Autonomous UGVs (laboratory-scale mobile robot Pioneer3DX) in a convoy formation outside Infolab21, Lancaster University campus

autonomous system must have the ability to learn quickly (from a single or few data samples) and to extract knowledge from the data streams collected by the sensors in real time, to rank the previously existing knowledge and to compare the relevance of the new knowledge to the previous knowledge leading to an update of the world model. The role of specific types of machine learning that are particularly suitable for online, real-time update of a real world model with evolving (growing or shrinking) nature is vitally important for the development of truly autonomous systems.

1.3 System Identification – an Abstract Model of the Real World

An autonomous system must have a model of the world (the environment that surrounds the AS and its internal functioning). Usually, this model is in the context of the goal that the AS must perform. Development of such models is governed by the system identification (Ljung, 1987), which is a topic usually considered in relation to control theory. Systems are often considered to be described by a set of (differential) equations. Alternative representations, for example statistical Bayesian, Markov models, decision trees, and so on. are also viable world models (see Chapter 2). An alternative that is particularly suitable to represent intelligent systems and knowledge is the fuzzy rule-based form of representation (to be discussed in Chapter 4). Whichever framework is used, however, the identification is usually considered in terms of:

a. the structure (with heavy human involvement, usually offline, at the design stage); and
b. parameters (often automatically, online, during the process of exploitation).

In what follows the concept for each of the two key aspects of identification problem will be briefly described.

1.3.1 System Structure Identification

The structure of the world model or the system is usually considered to be suggested by the human expert. It may take the form of:

a. a set of differential equations;
b. transfer function (time or frequency domain);
c. a set of fuzzy rules;
d. a neural network;
e. a stochastic model (e.g. Markov states model), and so on.

In this book, without limiting the concepts, the last three forms will be considered as examples. The main reason is their suitability to represent human intelligible knowledge in a granulated form.

System structure in the case of differential equations may comprise of the number and type of the differential (or difference) equations, the number of inputs and outputs. In the case of the transfer function it may include the order and type (e.g. 'all poles' or 'all zeros'). For neural networks (NN) the structure may define layers, feedback or feedforward, memory, number of inputs, outputs, and other elements that are optional. For the case of fuzzy systems the structure includes the following:

a. number of fuzzy rules;
b. number of inputs (features) and outputs;
c. type of the membership functions and their position in the data/feature space (this is not necessary for the specific type of fuzzy rule-based models considered in Section 4.3);
d. type of antecedents (scalar or vector);
e. type of the consequents (e.g. Zadeh–Mamdani (Zadeh, 1975; Mamdani and Assilian, 1975) or Takagi–Sugeno (Takagi and Sugeno, 1985));
f. type of connectives used (AND, OR, NOT);
g. type of inference (centre of gravity, 'winner takes all', etc.).

These will be further detailed and described in the next chapter.

Structure identification is an open research problem that still does not have a satisfactory and universally accepted answer. Structure identification can be seen as a nonlinear optimisation problem (Angelov, Lughofer and Klement, 2005) that aims to select the best structure in terms of minimum error in prediction/classification/control. Usually, it is not solved directly, but the structure is assumed to be provided. In some works the authors apply genetic algorithms, GA (Michalewicz, 1996), genetic programming, GP (Koza, 1992) and other numerical techniques for (partially) solving it. In this book a systematic approach will be used that is based on density increment that relates to the data density and distribution in the data space also taking into account the time element (*shift* of the data density). A fully theoretical solution is difficult, if possible at all.

Figure 1.4 illustrates in a very simplistic form the difference between the proposed and the traditional approach with respect to the role of the system structure identification – in ALS it is part of the automated process, while traditionally it is outside of the loop of automation.

There are different ways to devise automatically the structure of the model, including data space (regular) partitioning (Carse, Fogarty and Munro, 1996), clustering (offline or online and evolving) (Chiu, 1994; Babuska, 1998; Angelov, 2004a), based on data density (Angelov, 2002), based on the error (Leng, McGuinty and Prasad, 2005), and so on. The principle behind most of them is the old Latin proverb '*divide et impera*' which means 'divide and conquer' and leads to decomposition of a complex problem into (possibly overlapping and interdependent) subproblems or subspaces of the data space. The key questions that arise are:

- How to divide the problem or data space objectively (based on data density or the error are two obvious options); note that the traditional criteria for cluster quality, for example (Akaike, 1970) and so on, are designed to separate clusters well while

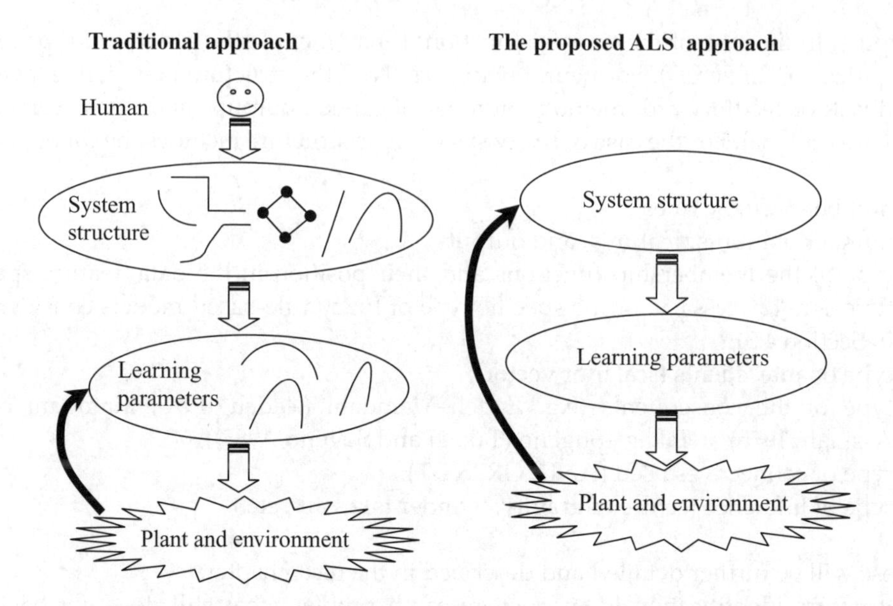

Figure 1.4 The traditional versus the proposed approach

for the purpose of model structure identification the overlap must be tolerated to avoid abrupt transitions between local models and gaps between them.

- Shall a data sample that is an outlier (which differs significantly from the existing local models or clusters) be ignored or it may be a start of a new local model (regime of operation); this problem is much more acute in online and real-time implementations when the decision must be taken based on the current data samples and no or little history.
- The optimality of the structure is, generally, a nonlinear problem, and therefore, its solution is, in principle, possible only numerically and offline; a possible pragmatic solution is to optimise the parameters subject to a structure that is selected automatically, but the optimality is then conditioned on the assumptions (as in other existing approaches).
- The dilemma between plasticity and stability – how often the structure can and should change – if it changes too often the system will lose its robustness; if it changes very rarely it will lose its sensitivity.
- Ideally, an automated algorithm for model structure identification should not depend on user- or problem-specific thresholds and parameters.

1.3.2 Parameter Identification

The problem of parameter identification is a much more established one (Ljung, 1987). The aim is to determine the optimal values of parameters of the model/system in terms of minimisation of the error of prediction/classification/control. If we use a fuzzy rule-based model as a framework that include parameters of the consequents of the fuzzy rules and parameters of the membership functions of the antecedent part of the rules (to be described and discussed in Chapter 4). If we use the particular type of fuzzy rule-based model introduced recently by (Angelov and Yager, 2012), (see Section 4.3) then the antecedent part is nonparametric.

The problem of parameter identification is also an optimisation one, but often can be considered as a linear or quadratic optimisation that guarantees uniqueness of the solutions subject to certain constraints. This is the basis of the widely used recursive least squares (RLS) method (Ljung, 1987). For general, nonlinear cases, they also use numerical procedures, such as error back-propagation, EBP (Werbos, 1990), other gradient-based techniques (e.g. conjugate gradients approach), and so on.

1.3.3 Novelty Detection, Outliers and the Link to Structure Innovation

The topic of novelty (respectively, anomaly) detection is pivotal for fault detection (Filev and Tseng, 2006) and video-analytics (Elgammal *et al.*, 2002; Cheung and Kamath, 2004). It has its roots in statistical analysis (Hastie, Tibshirani and Friedman, 2001) and analysis of the probability density distribution. The rationale is that novelties (respectively, anomalies, outliers) significantly differ and their probability density is significantly lower. Therefore, the test for a data sample to be considered as an outlier/anomalous is to have a low density.

The problem of system structure identification, especially in real time, is closely related to the outliers and anomaly detection, because an outlier at a given moment in time may be a start of a new regime of operation or new local model. In such case, the structure innovation will lead to increase of the density locally (around the new focal point). In this book we argue that the data density (local and global) can be used as an indicator for automatic system structure innovation and identification. *A drop of the global density indicates an innovation; an increase of local density indicates a consolidation* of a new regime of operation/new local behaviour.

1.4 Online versus Offline Identification

Autonomous systems have to be able to process and extract knowledge from streaming data in a so-called *online* mode. This means that the data stream is being processed sample-by-sample (here sample also means data item/instant) in a serial fashion, that is, in the same order as the data item was fed to the ALS without having the entire data stream/set available from the start. Imagine, a video stream – online processing (Figure 1.5, right) means processing it frame by frame, not storing (buffering) the whole video and then processing it *offline* (Ramezani *et al.*, 2008).

Systems that operate in offline mode may be good in scenarios that are very close or similar to the ones that they are specifically designed and tuned for. They need, however, to be redesigned or at least retrained/recalibrated each time when the environment or the system itself changes (e.g. in industry, the quality of raw materials, such as crude oil entering a refinery varies; catalysers are being removed or added to

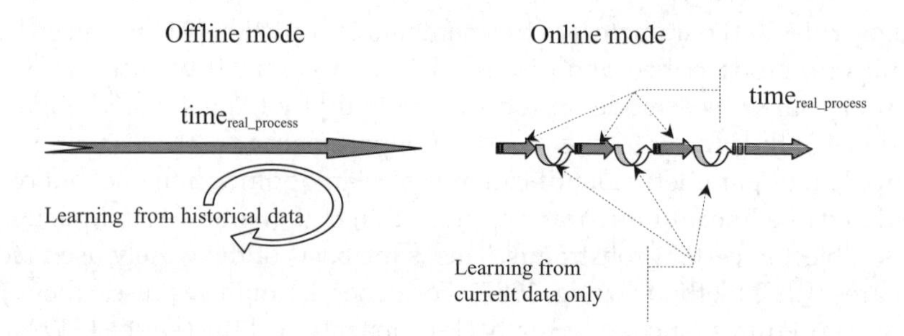

Figure 1.5 Online and offline modes of operation of a system

the polymerisation tank; hackers change their behaviour when they attack a computer system, UGV may enter an unknown zone, faults may develop in the subsystems of an AS, etc.). Offline systems (Figure 1.5, left) work with a historical 'snapshot' of the data *stream* and require all the previous data, which implies a much higher memory and computational requirements. In contrast to that, online systems work on a per sample basis and only require the current data sample plus a small amount of aggregated information; they do not require all the history (all previously seen data samples) of the data stream.

The online mode is often related to the real-time operation. It is important to stress that there is a subtle difference in the sense that an algorithm can be online (in terms of not storing the whole data sequence and processing data items one-by-one) and yet it might work slowly enough to be real time (if the real-life process is very fast while the computer processing unit, CPU is not that fast). In such cases there will be a delay in the output (prediction, class label, control action) produced by the model/system with respect to the real-world response. At the same time, a system may operate in a *real-time* manner and yet be offline if the sampling rate is extremely low. For example, in some biomedical problems the sampling (frequency of visits to the doctor and taking of measurements) can be as low as once every week or even month. In such cases, the system can learn from the whole previous history, process all previous data samples iteratively. Since these are extreme cases, in this book the focus will be on the ability of autonomous systems to learn online and in real time. Moreover, we will be primarily interested in so-called *recursive* algorithms that assume no iterations over past data, no storage/buffering of previous data and in a so-called *one-pass* processing mode.

1.5 Adaptive and Evolving Systems

As was already said, an AS should adapt to the environment. The theory of adaptive systems (Astroem and Wittenmark, 1989) is now a well-established part of control theory and digital signal processing (Haykin, 2002). It usually is restricted to systems with linear structures only and, more importantly; it does not consider the problem

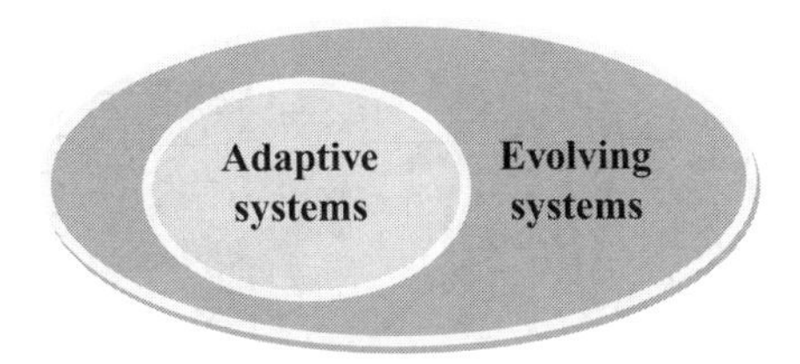

Figure 1.6 Evolving systems as a superset of adaptive systems

of system structure adaptation. An adaptive system is considered a system with a *fixed*, known structure that allows its parameters to vary/be adjusted. In this respect the concept of evolving systems (Angelov, 2002) as a system with evolving structure differs significantly. It is true, however, that evolving systems are also adaptive, but the subject of the adaptation are both system parameters as with the adaptive (in a narrow sense) systems as well as its structure. In this context, evolving systems can be seen as a superset of adaptive systems, Figure 1.6.

The area of evolving systems (as described above) that was conceived around the turn of the century (Angelov and Buswell, 2001; Angelov, 2002; Kasabov and Song, 2002) is still under intensive development and 'fermentation'. It is closely related to (albeit developing independently from) the works on self-organising systems (Lin, Lin and Shen, 2001; Juang and Lin, 1999) and growing neural networks (Fritzke, 1994). In the late 1990s and until 2001–2002 the term 'evolving' was also used in a different context – in terms of evolutionary (this will be clarified in the next section). Since 2002 and especially since 2006 when the IEEE started supporting regular annual conferences and other events (the last one, the 2012 IEEE Conference on Evolving and Adaptive Intelligent Systems, being in May 2012 in Madrid) it is used for dynamically evolving in terms of system structure systems. In 2010, the publishing company Springer started a new journal on Evolving Systems (http://www.springer.com/physics/complexity/journal/12530) and the number of papers and citations is growing exponentially.

The research area of evolving systems is central to the very notion of autonomous systems and autonomous learning and this will be made clearer and detailed in the rest of the book.

1.6 Evolving or Evolutionary Systems

In computational intelligence research evolutionary algorithms (EA), including such specific examples as genetic algorithms, GA (Goldberg, 1989; Michalewicz, 1996), genetic programming, GP (Koza, 1992), artificial immune system (Kephart, 1994), and so on. are computational algorithms that borrow heavily from the natural evolution. They often use a 'directed' random search for solving loosely formulated optimisation problems. They mimic a specific aspect of the natural evolution that is related to the population-based genetic evolution that is driven by such mechanisms as *mutation*,

Figure 1.7 Human beings are a good example of an ALS that evolves by learning (new rules) from experience through their sensors using their brain

chromosomal *crossover, reproduction, selection.* The natural evolution also has the aspect of individual self-development, especially for the case of human beings (Figure 1.7). Starting as small babies we do not have any idea about the surrounding world, but we start to collect data streams through our sensors and soon we start to create rules using and evolving our brains. We start to recognise what is *good* and what is *bad*, what is *dangerous* and what is *safe*, and so on. With time, our rule base grows; some rules stop being used or become irrelevant or need some adaptation and adjustment throughout our whole life. Some rules we are taught, some we infer ourselves.

It is interesting to note that the rules we acquire, update or stop using are not precise, for example 'IF *we lift a bag weighing over 63.241 kg* THEN *we will get a broken back*', but they are rather ***fuzzy***, for example. 'IF *we lift* ***heavy*** *loads* THEN *we may get a broken back*' or 'IF *it is cold* THEN we *take a coat*', and so on.

In essence, we self-develop. In this book we propose a systematic approach that allows building autonomous systems with such capabilities – to self-develop, to learn from the interaction with the environment and through exploration.

The Oxford Dictionary (Hornby, 1974, p. 358) gives the following definition of **genetic** – "a branch of biology dealing with the heredity, the ways, in which characteristics are passed on from parents to off-springs". The definition of **evolving** it gives (p. 294) is "unfolding; developing; being developed, naturally and gradually". In brief, despite some similarity in the names, EA differ significantly from the more recently introduced concept of evolving systems. While genetic/evolutionary is related to **populations** of individuals and parents-to-offspring heredity, evolving is applicable to **individual** system **self-development** (known in humans as autonomous mental development, Figure 1.7). 'Evolving' relates more to learning from experience, gradual change, knowledge generation from routine operation, rules extraction from the data. Such capabilities are vital for autonomous systems and, therefore, we will expand this idea in the book.

If we consider a fuzzy rule-based system as a framework, an *evolving* FRB system will learn new rules from new data *gradually* preserving majority of the rules learned already (Angelov, 2002). This is very similar to the way that individual people learn, see Figure 1.7. In a similar way to humans, an evolving fuzzy system (EFS) can be initiated by an initial rule base (in a supervised manner the way we learn from parents and teachers) or can start learning 'from scratch', autonomously.

1.7 Supervised versus Unsupervised Learning

The very notion of autonomous systems is closely related to the unsupervised learning and reinforcement learning (Sutton and Barto, 1999). However, semisupervised learning also has an important part to play because pragmatically no autonomous system is assumed to be reproductive and out of users' (human's) control in terms of monitoring – remember the famous Azimov's laws of robotics (Azimov, 1950). In other words, the level of autonomy of systems that are of practical interests for industry, including defence and security is 4 or maximum 5a according to Table 1.1. Examples of systems with lower level of autonomy (1–3) are decision support systems, DSS (McDonald, Xydeas and Angelov, 2008).

In this book, we are interested in autonomy of the knowledge extraction from data streams (autonomous learning) that (the same as the overall scheme of an AS, Figure 1.2 – see the smiley face at the very top of the figure) does not fully exclude the human user, but reduces his/her role to bare provision of goals and monitoring plus the option to abort the operation on safety grounds (autonomy level 5a, see Table 1.1). Provision of goals itself can be a source of definition of criteria for optimisation and learning objectives. Most often the latter are related to minimisation of the prediction error, maximisation of the classification rate, and so on.

The autonomous learning (AL) that can enable AS to adapt and evolve should acquire more than a simple input–output mapping that is typical for traditional (machine learning, fuzzy systems, neural networks, etc.) model learning techniques.

Table 1.1 Autonomy levels adapted from (Hill, Crazer, and Wilkinson, 2007)

Level	Autonomy	Authority	Interaction
5b	Full	Machine **monitored** by human	Machine does everything autonomously
5a			Machine chooses action, performs it and informs human
4b	Action unless revoked	Machine **backed up** by human	Machine chooses action and performs it unless human disapproves
4a			Machine chooses action and performs it if human approves
3	Advise and, if authorised, act	Human **backed up** by machine	Machine suggests options and proposes one of them
2	Advice	Human **assisted by** a machine	Machine suggests options to human
1	Advise only if requested	As above **when requested**	Human asks machine to propose actions and human selects
0	None	Human	Whole task done by human except for actual operation

Instead, the emphasis in AL is on building and constantly monitoring the quality and updating the structure of the system. The extracted knowledge usually (but not necessarily) is in the form of human interpretable, fuzzy rules (Hopner and Klawonn, 2000). This learning is 'on the fly' starting from few or even a single data sample, if needed, adapting quickly, but also being able to accommodate previous knowledge (if it exists) and fuse it with the newly acquired knowledge.

The most effective scheme proved to be the combination of unsupervised learning for model structure identification and semisupervised learning for parameter adjustment where the supervision comes from the data stream but with a time delay and not necessarily after each time step. The key in this scheme that is very much like the scheme of adaptive filtering (Haykin, 2002) and adaptive control (Astroem and Wittenmark, 1989) is the timing. The data stream often provides both the input and output in terms of the AS but at the moment of prediction/classification/control action generation a value can be unavailable (thus, the need to be predicted) while at the next time instant (see Figure 1.5) these values (if available and measured) can serve to feed back the learning in a supervised form. In this way, the supervised learning can also be considered as an automatic process that is related more to the online form of operation.

For example, if a system automatically models/infers/predicts the value of the outside temperature tomorrow or the exchange rate tomorrow based on some measurements and previous observations (history) then these predictions will be very useful until we get the real/true value the next day (so, in some 23–24 hours we can benefit from these predictions). The next day, we can use, however, the real/true values (if they are available because it may be available only sometimes, not necessarily every day). If and when the true values are available an autonomous learning system will be able to adjust and evolve without any direct human intervention.

1.8 Structure of the Book

The book is structured in three main parts preceded by this introductory chapter and closed by an Epilogue. This introductory chapter provided the motivation, background, a brief review of the previous and existing research work and publications in related areas as well as sets up some of the basic terminological definitions in the context of ALS.

The first part is dedicated to the systematic foundations of the methodology on which the ALS is based, including basics of probability theory (Chapter 2), pattern recognition and machine learning and especially clustering and classification (Chapter 3), the basics of fuzzy systems theory including neurofuzzy systems (Chapter 4).

Part II describes the methodology of autonomous learning systems. Chapter 5 introduces the *evolving* systems covering topics like data space partitioning, proximity measures, clustering, online input variable selection, monitoring the quality, utility and age of clusters, and so on. Chapter 6 describes the methodology for autonomous learning of the parameters of *evolving* systems stressing the difference between local and global learning methods. It also describes multi-input–multi-output (MIMO) systems, the inference mechanisms and methods for autonomous normalisation and standardisation of the data streams that the ALS processes online. In this chapter the fuzzily weighted recursive least squares (wRLS) method is described in the context of various possible learning modes. The issue of outliers and *drift* are discussed in the context of robustness.

In Chapter 7, the autonomous predictors, estimators, and filters are described. They are powerful tools for addressing time-series modelling and a range of other related problems of adaptive estimation and filtering. For example, the methodology behind one of the very interesting applications of ALS – autonomous sensors, AutoSense, is described in more detail in this chapter form the theoretical point of view and is revisited in Part III of the book from the application point of view.

Chapter 8 describes the autonomous classifiers using AutoClassify as an example that is based on evolving clustering and fuzzy rule-based systems.

Chapter 9 outlines autonomous learning controllers, AutoControl based on the concept of evolving fuzzy rule-base and the relatively old concept of indirect adaptive control.

Finally, Chapter 10 closes Part II with a discussion of collaborative ALS – a topic that has large potential for future development mainly in robotics, defence and related areas of security, surveillance, aerospace, and so on.

Finally, Part III is dedicated to various applications of the ALS with the clear understanding that the list of applications that the author and his students and collaborators have developed during the last decade is open for expansion. Indeed, a growing number of publications by other authors in the area of *evolving*, autonomous learning system, the regular IEEE conferences and events on this topic illustrate the huge potential for further growth. The adoption by leading industrial companies of these ideas demonstrates the potential which these pioneering concepts have for the Economy and the Society.

Chapter 11 describes the application aspects of AutoSense to a range of products (e.g. kerosene, gasoil, naphta) of a real large-scale oil refinery located in Santa Cruz de Tenerife, owned and run by CEPSA, Spain. One particular problem discussed in this chapter that has safety implications is the autonomous prediction of inflammability index (e.g. Pensky-Martens or Abel (Ishida and Iwama, 1984)) in real time. In the same chapter another range of application studies (courtesy of Dr. Arthur Kordon, The Dow Chemical, Texas, USA) are described. These include chemical compositions and propylene.

Chapter 12 is focused on the application issues of AutoClassify and AutoCluster in mobile robotics. The illustrative examples and video material are available at www.wiley.com/go/angelov. Both landmark identification and recognition and navigation and control subtasks were considered.

Chapter 13 describes applications of the recursive density estimation (RDE) approach to video surveillance applications (autonomous novelty detection and object tracking in video streams) that the author and his students introduced recently.

Chapter 14 provides a description of the application of the proposed ALS approach to model *evolving* user behaviour. This applies to users of computers, home appliances, the Internet, and so on. Most of the existing approaches ignore the aspect of dynamic evolution of the behaviours of the users and considers them as 'averaged' statistics very much in the sense of 'one size fits all' paradigm. The proposed ALS approach allows personalisation and learning specific users) in real time.

The book also provides a source of basic mathematical foundations used in the text, discusses the problems of real-life applications and will be very useful to be used with the software package available at www.entelsensys.com.

Additional teaching material (slides) that can be used for short courses or lectures can also be downloaded from the above website.

PART I

Fundamentals

2

Fundamentals of Probability Theory

It is a truth very certain that when it is not in our power to determine what is true we ought to follow what is most probable.

(R. Descartes in Discourse on Method)

I will never believe that God plays dice with the Universe.

(A. Einstein)

Probability theory is one of the methodologies to represent and tackle some types of uncertainties (specifically, randomness). It was mainly developed in the eighteenth century with main contributions from the mathematicians such as Blaise Pascal (1623–1662), Pierre de Fermat (1601–1665), Daniel Bernoulli (1700–1782) and later the British clergymen Thomas Bayes (1701–1761) addressing problems of gambling and insurance. A huge role to make it more mathematics based and scientific was played by the Russian mathematician Andrey A. Markov (1856–1922) and the Soviet academic Andrey N. Kolmogorov (1903–1987).

This theory is widely used and read in most prestigious universities. It is the view of the author of this book that it indeed provides an elegant framework to describe random processes and data, but is overused, because the assumptions on which it is based (random nature of the processes and events, independence of the data samples from one another and, often required, normal or parameterised distributions) rarely (or, more precisely, never fully) hold in practice. A possible explanation why probability theory is accepted much more widely than, for example, fuzzy logic theory is, perhaps, its philosophical and attitude/mentality closeness to the opportunism (taking chances) and gambling rather than to determinism or to dialectics – something

Autonomous Learning Systems: From Data Streams to Knowledge in Real-time, First Edition. Plamen Angelov.

that is much more dominant in the so-called Anglo-Saxon world that itself is dominant now – a good example is the economic and financial crash of 2008.

Nevertheless, a brief introduction to probability theory will be provided here for the following reasons:

a. In real processes (mainly because of the complexity of the underlying physical, biological, economic, etc. phenomena) there are components that are not described fully that leads to so-called *noise* that, indeed, can be considered as a random in nature data stream.
b. Probability theory is closely related to the statistical analysis – so-called *frequentistic* approach (Hastie, Tibshirani and Friedman, 2001) that dominated over the more recent *belief-based* approach (Demspter, 1968) that considers the probability from an epistemiological point of view; thus, it plays an important role of a bridge between the statistics and fuzzy logic (Liu and Yager, 2008).
c. There is some duality between the main results of the theory of probability and the results in other areas such as neural networks, fuzzy systems, machine learning, which will be stressed further.

2.1 Randomness and Determinism

There are several problems that are of interest for ALS and will be considered in this book, such as:

i. clustering (grouping the data);
ii. classification (supervised clustering with labels for the classes);
iii. prediction, estimation, filtering (time series, prognostics, regression);
iv. control (adaptive, self-learning controllers);
v. outliers (anomaly/novelty) detection;
vi. automatic inputs selection (sensitivity analysis);
vii. collaboration between more than one ALS.

The philosophical question that most of these problems, which come from the areas of control, machine learning and pattern recognition theories, raise can be formulated as '*can we make* ***valid*** *assumptions for future values, distributions, class labels or number of clusters based on some (past and present) observations?*'.

The short (Bayesian) answer to this question is 'we make *a priori* estimation that we update once *a posteriori* information is available'. They use Bayes theorem to make this inference (Hastie, Tibshirani and Friedman, 2001):

$$p(Y \mid X) = \frac{p(X \mid Y)p(Y)}{p(X)} \tag{2.1}$$

where

$p(.)$ denotes a probability density function;
$p(.\,|\,.)$ denotes the conditional probability;
X denotes *prior* and Y – the *posterior;*
$p(Y \mid X)$ denotes the probability that Y will take a value y if X takes value x.

This concerns the probability density function (pdf) that will be discussed in Section 2.3 and relies heavily on the strong assumptions regarding data distributions, (in-)dependency, and random nature.

The data mining and machine learning short answer to the same question is to solve an optimisation problem which has to provide the 'optimal' clusters/classifier/predictor/controller/estimator/filter. Maximum likelihood optimisation leads to analytical solutions only in a limited circle of problems when strong assumptions regarding the data distributions and (non)linearity are made (Bishop, 2009).

The (adaptive) control theory short answer to the same question is to perform an 'estimate–update' pair of actions similarly to the *prior–posterior* pair at each time step (Astroem and Wittenmark, 1989). In this book we also adopt the adaptation approach and combine it with the use of statistical information regarding the data density that is not the same as (although is similar to) probability density functions; the main difference is that it does not integrate to 1. It also does not require unlimited observations ($N \to \infty$) while probability theory applies for large numbers only. There is also no need for the observations to be independent or to know their mutual dependencies apart from their mutual position in the data space (closeness).

For example, the two main results of the frequentistic approach are the so-called central limit theorem and the strong law of large numbers (Bishop, 2009). Both require (are correct for) an infinitely large number of observations ($N \to \infty$). Therefore, the use of the probability theory must be done with care and with attention to the assumptions on which the principles, laws and conclusions of this theory are based upon. For example, based on a single or a small number of observations one can not apply the laws of probability theory with a solid justification.

In this book an innovative approach is presented that is a more deterministic solution in which the uncertainties are considered as a structural phenomenon (the structure does not fully represent the complexity of the problem and we allow it to evolve) rather than to be considered as (purely) independent, random one. They are addressed by adaptation (which concerns not only the parameters as in adaptive control theory), but at the same time, it is ***predictable***, because for the same data stream (including the order of the data) the result will *always be the same*, unlike random systems.

The author of the book strongly believes that this is more appropriate form of solution of the problem of stability–plasticity dilemma that addresses the question *"how learning can be processed in response to significant input patterns and at the same time not to lose the stability for irrelevant patterns"* (Carpenter and Grossberg, 2003).

Indeed, there are processes in Nature that are 'purely'/strictly random (such as gambling games, etc.), but many other processes that we actually want to describe, control or classify are not random and the elements of 'randomness' that we observe are, in fact, a result of our limitations in describing and handling them. For example, climate is not random by nature; it is, rather, too complex to describe completely; the same is true, for example, for the economic data, social and biological processes, and so on.

If the complexity of a problem is high, an old principle used over the centuries is '*divide et impera*' – to decompose the problem into a simpler set of subproblems; another useful approach is to consider adaptation and evolution as a way to adapt a simple solution to a more complex and dynamic phenomena. These are precisely the two pivotal principles (*divide et impera* and adaptation and evolution) that underpin the theory of *evolving* and autonomous learning systems.

2.2 Frequentistic versus Belief-Based Approach

The *frequentistic* approach is the older one. It considers the probability as a frequency of occurrence of an event. For example, if it was raining during 10 of the days of a month (assuming months with 30 days) then the probability of raining is said to be $p = 0.3(3)$ (33(3)%) or 1/3):

$$p(X) = \frac{N_x}{N} \tag{2.2}$$

where

N_x denotes the number of rainy days;
N denotes the number of all days in the month;
X in this case is the event that the day is *rainy*.

The belief (or evidence) approach to probabilities is closer to the betting and the degree of belief that an event (for example that it will rain tomorrow) will take place.

Obviously, the belief-based approach is more subjective, epistemiological, while the *frequentistic* approach is rooted in statistics. Although the belief-based approach to probabilities has closer links with fuzzy logic theory (Liu and Yager, 2008) some elements of which we will use, we prefer and will stick to the *frequentistic* version of the probability theory. The reason is that it is more objective and data-centred; in a similar way, the elements of fuzzy logic (FL) that we will use are related to basic concepts such as *partial degree of membership* that can be elicited and represented by the data distribution and are, thus, objective. This is somewhat different form the traditional subjective (expert-based or related) character of the FL.

2.3 Probability Densities and Moments

Probabilities are non-negative (since they represent frequencies) and are represented by probability densities that are positive values from the range [0;1] normalised to sum up to 1:

$$0 \leq p(x) \leq 1 \tag{2.3}$$

$$\int_{-\infty}^{\infty} p(x)dx = 1 \tag{2.4}$$

where x denotes the value that the random variable X can take.

The second condition follows from the definition of the probability that the value of x will be in certain interval/range (a, b):

$$p\left(x \in (a, b)\right) = \int_{a}^{b} p(x)dx \tag{2.5}$$

Weighted averages play an important role in the calculus (Yager, 1988). Probabilities are used as weights in the estimation theory and lead to the definition of *expectations*:

$$E(f(x)) = \int_{-\infty}^{\infty} p(x)f(x)dx \tag{2.6}$$

When the data are finite (taken from a data stream or set) expectation can be represented as a sum and that is called a *mean* or *average*:

$$E(f(x)) \approx \frac{1}{k}\sum_{i=1}^{k} p(x_i)f(x_i) \tag{2.7}$$

In this book, when referring to statistical (frequentistic) mean/average the following notation will be used:

$$\mu_k = \frac{1}{k}\sum_{i=1}^{k} p\left(x_i\right) f\left(x_i\right) \tag{2.8}$$

In a vector form of a (first) norm it can also be written as:

$$\mu = \|p\left(x\right) f(x)\| \tag{2.9}$$

Expectation (or mean) plays a very important role in statistics and machine learning. It is called the first moment. Because the random variables and processes can not be described with certainty, they can (only) be described 'on average', as an estimate. The traditional mean as described above, however, has the following problem: in some cases it may represent an infeasible point.

For, example, let as consider a simple example of throwing fair die. The expected value (the mean) is actually an infeasible value of 3.5, because:

$$\mu = \frac{1}{6}\sum_{i=1}^{6} i = \frac{1}{6} + \frac{2}{6} + \frac{3}{6} + \frac{4}{6} + \frac{5}{6} + \frac{6}{6} = 3.5$$

Obviously, the mean is only one of the forms of representing the expectation for a random variable or process and is not enough and moreover, may even have infeasible value! Another important moment (of second order) is the *variance*, which provides a statistical measure of the deviation from the mean:

$$\mathrm{var}[f] = E[(f(x) - E[f(x)])^2] \tag{2.10}$$

The notation that will be used in this book has statistical/frequentistic sense:

$$\sigma^2 = \|f(x) - \mu\|^2 \tag{2.11}$$

Its square root, σ is also-called the standard deviation. Strictly speaking, the expression (2.11) is *biased* in terms of maximum likelihood and expectation. The unbiased expression assuming k values in total is given by:

$$\sigma^2 = \frac{1}{k-1}\sum_{i=1}^{k}(f(x) - \mu)^2 \tag{2.12}$$

When the variance concerns two or more independent random variables the covariance is defined in a similar way (e.g. for two variables, x and y):

$$\mathrm{cov}[x, y] = E_{x.y}[(f(x) - E[f(x)])(f(y) - E[f(y)])] \tag{2.13}$$

The notation that will be used in the book has again a statistical/frequentistic sense:

$$\Sigma_{x,y} = (f(x) - \mu_x)(f(y) - \mu_y)^T \tag{2.14}$$

The variance itself is not enough to represent the random variable or process because if the variance is high the information that the mean itself brings is less certain.

The density distribution of real data sets and streams is usually complex, multi-modal and variable with time, but in practice they often consider several 'standard' types of probability density distributions because of their convenience and role in theoretical derivations on which the theory of probability is based. The most prominent of these probability density functions (pdf) is the Gaussian one named after Carl Friedrich Gauss (1777–1855), which is also-called *normal*:

$$p_G = \frac{1}{(2\pi\sigma^2)^{1/2}} e^{-(x-\mu)^2/2\sigma^2} \tag{2.15}$$

The above definition is over a single variable, x. In a similar way, the Gaussian (normal) distribution over n-dimensional vector $\mathbf{x}$ of continuous variable can be defined:

$$p_G = \frac{1}{(2\pi)^{n/2}|\Sigma|^{1/2}} e^{-\frac{(x-\mu)^T\Sigma^{-1}(x-\mu)}{2}} \tag{2.16}$$

where $|\Sigma|$ denotes the determinant of the covariance matrix, Σ.

Other 'standard' types of pdf include Cauchy named after Augustin Cauchy (1789–1857):

$$p_C = \frac{1}{\pi\gamma\left(1+\left(\frac{x-x_0}{\gamma}\right)^2\right)} \tag{2.17}$$

which for $x_0 = 0$ and $\gamma = 1$ simplifies to what is known as the standard Cauchy function:

$$p_C = \frac{1}{\pi(1+x^2)} \tag{2.18}$$

It is interesting to note that the two functions (Gaussian and Cauchy) are closely linked. They have very similar shape (with the Cauchy function being sharper in the centre and decreasing asymptotically towards zero slower than the Gaussian one).

It can even be shown that the Cauchy-type function (Figure 2.1) can be considered as a first-order Taylor-series approximation of the Gaussian one, because:

$$e^{-\frac{\|x-\mu\|^2}{2\sigma^2}} = \frac{1}{e^{\frac{\|x-\mu\|^2}{2\sigma^2}}} \approx \frac{1}{1+\frac{\|x-\mu\|^2}{2\sigma^2}+\ldots} \tag{2.19}$$

The main problem in reality is that the distributions of the real data sets/streams are usually quite different from the 'standardised' distributions that are, on the other

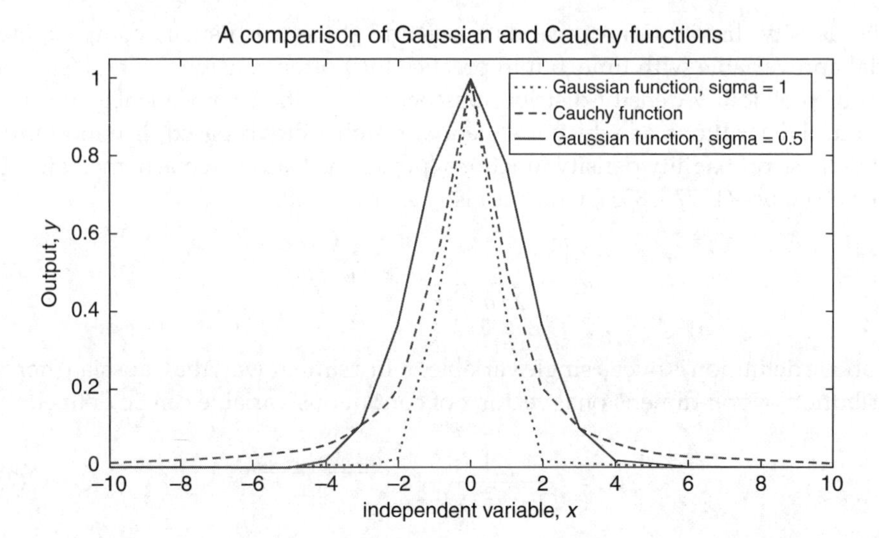

Figure 2.1 An example of Gaussian and a Cauchy type of functions

hand, very convenient and suitable for theoretical derivations. Real distributions are multimodal (have many local extreme points), not smooth and time varying (nonstationary). In this respect, the quote from Albert Einstein "as far as the laws of mathematics refer to reality, they are not certain; and as far as they are certain, they do not refer to reality", quoted in (Newman, 1956) is very appropriate.

2.4 Density Estimation – Kernel-Based Approach

Kernels became popular recently because they allow representing a multimodal (as opposed to the standard, unimodal) distribution as a set of simpler, kernel representations that are valid locally. In this sense, the kernel approach is of the type of *'divide et impera'* and, therefore, it helps reduce the complexity of the real problems. One of the first kernels represent the so-called Parzen windows, named after Emanuel Parzen (1929).

Kernel density estimation (KDE) is a generic approach where the conclusions apply to the whole data distribution but the kernels are drawn from a finite set of representative data samples. Kernels apply locally (to a region, for example, let us denote such a region containing x by $\Re$). The probability mass associated with this region can then be defined as an integral/sum:

$$P = \int_{\Re} p(x)dx \tag{2.20}$$

where P denotes the probability mass of the kernel over a region $\Re$.

If we assume additionally that the region is so small that the probability $p(x)$ is roughly constant over the region, we can express the probability density function as (Bishop, 2009):

$$p(x) = \frac{N_{\Re}}{kV} \tag{2.21}$$

where

$N_{\Re}$ denotes the number of points that lie inside $\Re$;
k denotes the number of data points;
V denotes the volume of the region $\Re$.

The same constraints/requirements apply to kernels as to pdf, in general, namely eqns. (2.3) and (2.4):

$$0 \leq K(x) \leq 1 \tag{2.22}$$

$$\int_{-\infty}^{\infty} K(x)dx = 1 \tag{2.23}$$

Given k independent and identically distributed random data points x_i, $i = 1, \ldots, k$ in an n-dimensional space R^n with an unknown density p, the multivariate KDE $\hat{p}(x)$ at $\mathbf{x}$ is given as (Bishop, 2009):

$$\hat{p}(x) = \frac{1}{kh^n} \sum_{i=1}^{k} K\left(\frac{x - x_i}{h}\right) \tag{2.24}$$

where $K(\cdot)$ is the kernel function that is symmetric but not necessarily positive and integrates to one, and $h > 0$ is the radius.

Again, there are different types of kernels, for example Gaussian, Cauchy, Epanechnikov. For example, the Epanechnikov type of kernel is defined as follows (Bishop, 2006):

$$K\left(\frac{x - x_i}{h}\right) = \begin{cases} \frac{1}{2} V_n^{-1} (n+2)\left(1 - \left\|\frac{x - x_i}{h}\right\|^2\right); & \left\|\frac{x - x_i}{h}\right\|^2 < 1 \\ 0 & \text{otherwise} \end{cases} \tag{2.25}$$

where V_n is the volume of the unit n-dimensional sphere.

The overall probability, when using kernels, is usually defined as an average over the number of data samples considered:

$$p(x) = \frac{1}{k}\sum_{i=1}^{k} K\left(\frac{x^* - x_i}{h}\right) \tag{2.26}$$

where x^* denotes the centre/focal point of the kernel.

KDE has a single parameter – the kernel radius, h and its choice is still one of the weaknesses of this, otherwise, very accurate approach. In addition, the physical units in which h is being measured also influence the result. It is true that some problem-independent (objective) principles can be applied in the choice of the kernel radius (the only parameter that needs to be specified) such as the expected number of kernels in a normalised or standardised data space. However, another major disadvantage of the KDE approach is that it is computationally expensive because it is inherently offline (a sum of exponentials, in the case of Gaussians, is not easy, if possible at all, to calculate recursively).

2.5 Recursive Density Estimation (RDE)

Data density (Figure 2.2) plays a very important role in model structure design, novelty/anomaly/outliers detection (respectively, fault detection) and other related problems, including collision detection (Angelov, Ramezani and Zhou, 2008), video-analytics and so-called landmark detection and identification (Zhou and Angelov,

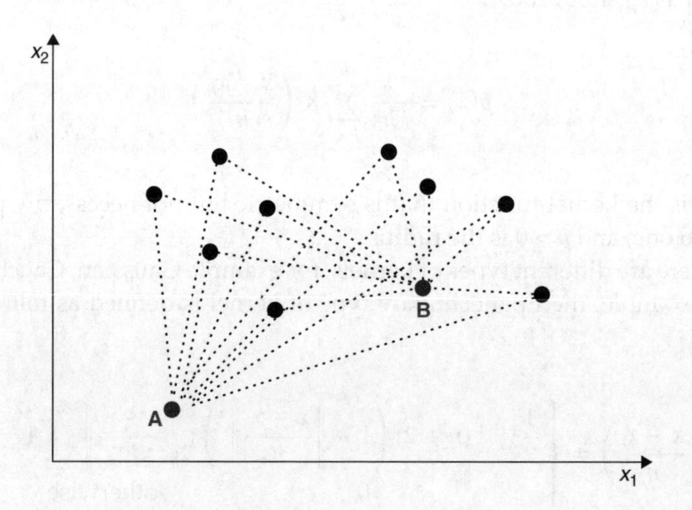

Figure 2.2 An illustration of the idea of density (to be distinguished from pdf). The density at point A is low while at point B is high

2007). Its usefulness, however, is greatly limited by its computational complexity and the requirement for offline calculations that impose limitations on both the memory and computational power (for large values of k). Additionally, the presence of the parameter, h also limits the universal applicability to different data streams.

To address these problems a recursive approach was introduced that dates back as far as 2001 (Angelov and Buswell, 2001), but got its name RDE in 2008 (Angelov *et al.*, 2008) and its latest version is a part of a patent application (Angelov, 2012). It is based on the Cauchy type of kernel (Angelov, 2004a):

$$D_k(x) = \frac{1}{1 + \overline{d}_k^2} \tag{2.27}$$

$$\overline{d}_k^2 = \frac{1}{k}\sum_{i=1}^{k} \|x_k - x_i\|^2$$

where D_k denotes the global density calculated at the moment in time, k; we will refer to it also in the future simply by D.

The mean norm, $\overline{d}_k^2$ is defined as the mean distance between the current data sample, x_k and all the other points seen so far. It can be of Euclidean form as shown in the second equation of (2.27), but equally, it can be of Mahalonobis or cosine or any other eligible type of proximities described in more detail in the next chapter.

The proposed form of density differs from the pdf, because, while condition (2.3) or (2.22) is satisfied, condition (2.4) or, respectively, (2.23) – is not. The value of the data density (by differ from the probability density) is equal to *1* when all the data samples have the same value that is also equal to the value of the focal point (a singular case that is rare in practice):

$$D_k(x) = 1 \quad \textit{iff} \quad x_i = x_k; \forall i \tag{2.28}$$

The proposed expression for the density is nonparametric (it does not have even the pretty generic parameter, h). It does depend on the relative data distribution while the standard probability density distribution assumes independence of the observations, which is true only for handful of problems like tossing coins, throwing die, lottery and other 'purely' random phenomena. In real processes the individual observations and measurements influence each other and Bayesian propagation rule (see Equation (2.1)) usually takes into account only two consecutive observations and cannot take into account ***all*** past measurements (this is the reason popular hidden Markov models, HMM are usually considered of so-called first order only). The data density expression (2.27) introduced by the author in 2001 takes into account ***all*** past observations and the relation of the current observation to them in terms of an inverse of the scatter measured from the current observation. While for typical 'random' processes such as tossing a coin, throwing die or playing a lottery the traditional, Bayesian inference and pdf formulation is appropriate and the author's view is that the data density formulation of the type of (2.27) is more appropriate for most of the real processes where we usually observe complexities that are difficult to describe rather than 'pure' random processes. The traditional pdf answers a different question as compare to the data density expression (2.27). The traditional pdf answers the question '*What is the probability that the value of the variable is ...?*' while the data-density expression (2.27) answers the question '*How the value ... of the variable relates to all previous/other values observed for this variable?*' that, in general and more often, is not the same. For detecting outliers, clustering, and complex system structure identification data density, D (not the traditional pdf, p) plays a critical role, as detailed later in the book.

Let us consider a very simple, yet illustrative, example to point the differences between the traditional pdf and the data density of the form of (2.27). Let us have a dice and let us have 3 on the dice the first time and try to estimate the probability of the event that the dice will have 6 following 3. Following the traditional Bayesian approach (2.1) one will get

$$p(6\,|\,3) = \frac{p(3\,|\,6)p(6)}{p(3)} = \frac{\left(\frac{1}{6}\right)\left(\frac{1}{6}\right)}{\frac{1}{6}} = \frac{1}{6}$$

The same will apply for any other value on the dice. If, however, the data density for the value 3 will be represented with a Cauchy type of curve with values as shown in Table 2.1 and Figure 2.3. For example, for the value 6 the density will be much lower than for the same value 3 and will be equal to:

$$D_2(6) = \frac{1}{1 + \frac{1}{2}\left((6-3)^2 + (3-3)^2\right)} = \frac{2}{11}$$

Note that the pdf and density, D address different questions as described above. The key moment is that for processes like tossing a coin, throwing die or lottery numbers prediction pdf and traditional Bayesian approach is more appropriate because all possible values are equally probable (since their sequence is 'purely' random). However, for most real processes there is **no** independence between ***all*** different observations/measurements, which is assumed in the Bayesian inference. The data density definition at (2.27) also does not require an infinite number of observations and can be more representative and useful, as will be described later in the book.

Table 2.1 A comparison of the traditional pdf with the data density distribution for the example of throwing a dice

x	1	2	3	4	5	6
pdf	1/6	1/6	1/6	1/6	1/6	1/6
D	1/3	2/3	1	2/3	1/3	2/11

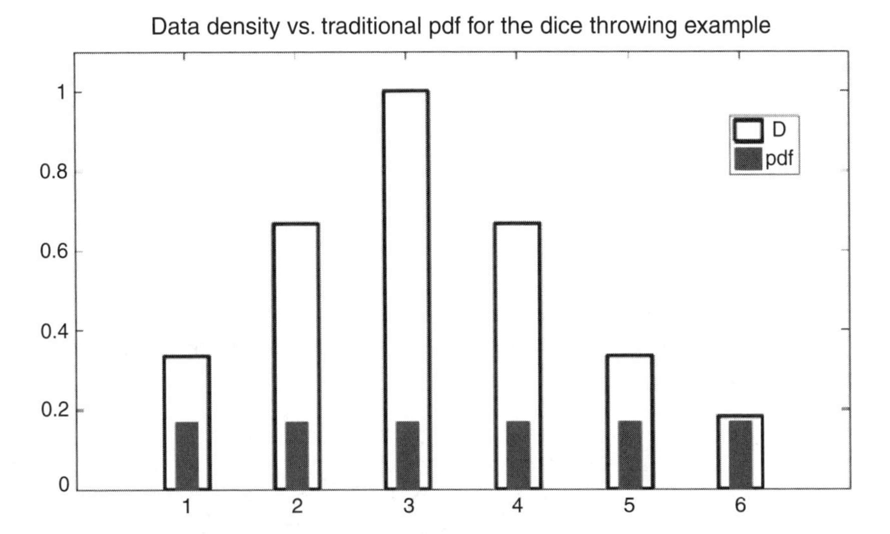

Figure 2.3 Traditional pdf versus the proposed data density distribution for the simple example of throwing a dice

Last but not least, the data density expression (2.27) unlike pdf makes possible recursive calculations (Angelov, 2004a; Angelov, 2012). This means that only a very small amount of data – only the mean of all data samples, μ_k and the scalar product averages quantity, $\sum_k$ calculated at the current moment in time k – are required to be stored in the memory and updated. The current data sample, x_k is also used, but it is available and there is no need to store or update it. This has huge implications, because it allows theoretically an infinite amount of data (infinitely large data sets or infinitely long and open-ended time-wise data streams) to be processed ***exactly*** (not approximately!) in real time, very fast. For example, a comparison was made for video-analytics problems where RDE outperformed the best-known KDE approach (Elgammal *et al.*, 2002) in orders of magnitude both time wise and computational complexity wise (Angelov *et al.*, 2011).

Local density (in addition to the global density) was also introduced on the basis of the kernel type of formulation for Euclidean type distance (Angelov and Filev, 2005):

$$d_{\Lambda}(x_k) = \frac{1}{1 + \frac{1}{N_{\Lambda}} \sum_{i=1}^{N_{\Lambda}} \|x_k - x_i\|^2} \tag{2.29}$$

where d_{Λ} denotes local density of region Λ; N_{Λ} denotes the number of data samples associated with the region Λ.

The recursive expression applicable to both global and local density has been derived as an ***exact*** (not approximated or learned) quantity as (Angelov, 2011):

$$D(x_k) = \frac{1}{1 + \|x_k - \mu_k\|^2 + \Sigma_k - \|\mu_k\|^2} \tag{2.30}$$

where both, the mean, μ_k and the scalar product, Σ_k can be updated recursively as follows:

$$\mu_k = \frac{k-1}{k}\mu_{k-1} + \frac{1}{k}x_k \quad \mu_1 = x_1 \tag{2.31}$$

$$\Sigma_k = \frac{k-1}{k}\Sigma_{k-1} + \frac{1}{k}\|x_k\|^2 \quad \Sigma_1 = \|x_1\|^2 \tag{2.32}$$

The recursive expressions of the RDE (2.30)–(2.32) are exact (they lead to exactly the same result as Equation (2.27)) and apply for both global (2.27) and local density (2.29). They can be derived, for example, starting from Equation (2.27). After simple reorganisation we get:

$$D_k = \frac{1}{1 + x_k^2 - 2x_k\frac{1}{k}\sum_{i=1}^{k} x_i + \frac{1}{k}\sum_{i=1}^{k} x_i^2} \tag{2.33}$$

This expression can be further regrouped into:

$$D_k = \frac{1}{1 + x_k^2 - 2x_k\frac{1}{k}\sum_{i=1}^{k} x_i + \left(\frac{1}{k}\sum_{i=1}^{k} x_i\right)^2 - \left(\frac{1}{k}\sum_{i=1}^{k} x_i\right)^2 + \frac{1}{k}\sum_{i=1}^{k} x_i^2} \tag{2.34}$$

This can be further simplified into:

$$D_k = \frac{1}{1 + \left\| x_k - \frac{1}{k}\sum_{i=1}^{k} x_i \right\|^2 + \frac{1}{k}\sum_{i=1}^{k} x_i^2 - \left(\frac{1}{k}\sum_{i=1}^{k} x_i\right)^2} \tag{2.35}$$

Using the following notations we arrive at Equation (2.30):

$$\mu_k = \frac{1}{k}\sum_{i=1}^{k} x_i \tag{2.36}$$

$$\Sigma_k = \frac{1}{k}\sum_{i=1}^{k} x_i^2 \tag{2.37}$$

2.6 Detecting Novelties/Anomalies/Outliers using RDE

There is a solid body of literature concerning statistical approaches for novelty (respectively, anomalies, outliers) detection, but the approaches are predominantly offline or require expert knowledge (Patton *et al.*, 2000). The problem from the machine learning point of view can be considered as a single-class classification. Moreover, they are unbalanced problems with the majority of samples being 'normal' and only a tiny minority of samples being 'anomalous'. Because outliers/novelties/anomalies are sparse they do not render multiclass classification problem as such. Well-established methods for addressing this problem include Hotteling statistics (Hotteling, 1931), cluster analysis (to be discussed in the next chapter) and more recent approaches such as one-class support vector machines (SVM) (Manevitz and Yousef, 2001) and Artificial Immune System (Krishnakumar, 2001).

One can group the methods for outliers/novelties/anomalies identifications as:

a. Fully unsupervised (when no *prior* information is assumed about the distribution of the data).
b. Model-based (both normal and abnormal distributions are being modelled using some form of a model – e.g. probabilistic, fuzzy, neural network, support vectors, etc.). Naturally, this approach requires active participation of the expert/supervisor. A natural subgroup of approaches considers only the normality be modelled.

There is no strict mathematical definition of what an outlier/anomaly is (Figure 2.4), but the so-called *3σ* principle (which follows from Chebyshev's theorem) is widely used.

Another widely used statistical principle relates to hypothesis testing and, for example, when comparing the deviation from multivariate Gaussian/normal distribution. It is often formulated by the so-called χ^2 principle that is based on precalculated tables with values for different dimensions of the input vector, n (Duda *et al.*, 2000; Hastie, Tibshirani and Friedman, 2001). In general, however, χ^2 tests assume normal

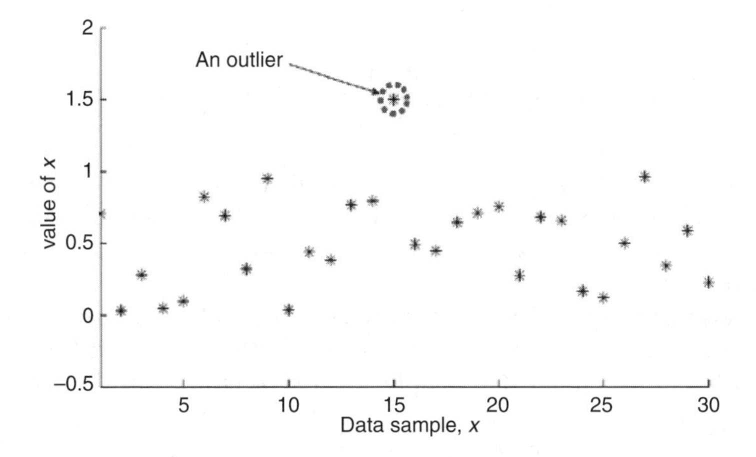

Figure 2.4 Outlier/anomaly – illustration

Gaussian distribution and are not recommended for high values of n (Georgieva and Filev, 2010).

Outlier detection is very important in fault, intrusion and fraud detection, video-analytics, cyber security, event detection in sensor networks, landmark detection, navigation, and, simply, data preprocessing (Duda *et al.*, 2000; Zhou and Angelov, 2007; Trevisan *et al.*, 2010).

Outlier/anomaly/novelty detection is a difficult task even if done offline (Hastie, Tibshirani and Friedman, 2001), but here an approach is offered that can be performed online. It is based on the RDE. Traditionally, statistical approaches are based on the density estimation. Based on the value of the density calculated by RDE a simple, yet efficient outlier detection method is proposed as follows (Angelov, 2012):

First, the density, D_k is calculated in real time per data sample as described earlier (2.30)–(2.32). The mean density, $\overline{D}_k$ is then:

$$\overline{D}_k = \frac{1}{k}\sum_{i=1}^{k} D_k \tag{2.38}$$

and it can also be updated in real time as follows:

$$\overline{D}_k = \frac{k-1}{k}\overline{D}_{k-1} + \frac{1}{k}D_k \quad \overline{D}_1 = D_1 \tag{2.39}$$

The variance of the density (not of the original signal, x) is then:

$$\left(\sigma_k^D\right)^2 = \left\|\overline{D}_k - D_k\right\|^2 \tag{2.40}$$

and it, too can be recursively updated by:

$$\left(\sigma_k^D\right)^2 = \frac{k-1}{k}\left(\sigma_{k-1}^D\right)^2 + \frac{1}{k}(D_k - \overline{D}_k)^2 \quad \left(\sigma_1^D\right)^2 = 0 \tag{2.41}$$

Based on Equation (2.39) and Equation (2.41), outliers are easy to detect and identify using the standard deviation, σ_k^D from the mean of the density, $\overline{D}_k$ which is illustrated in Figure 2.5 for an example of real-time video analytics (to be detailed in Chapter 13).

Another example (Figure 2.6), which is given here purely for illustrative purposes, is described in more detail in the report to the STAKE project with the UK MoD (2011), which is also mentioned in Chapter 12. It depicts an outlier in terms of phone calls made by a person who is *young* (21 years old), white. The density in terms of characteristics/features of these calls was particularly lower than for all the other callers. This was possible to be calculated automatically and in real time.

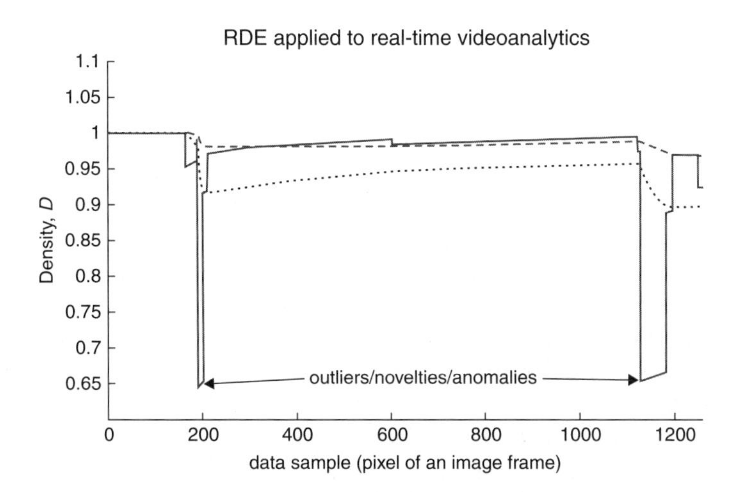

Figure 2.5 The density, D is plotted by the solid line and the mean is plotted by the dashed line, while the mean with subtracted standard deviation is plotted by the dotted line. The outliers/anomalies are clearly detectable by the sudden drop of the density. These outliers represent a new object (van) that appears on the scene observed by a security camera (a video-analytics application which will be described in more detail in Chapter 13)

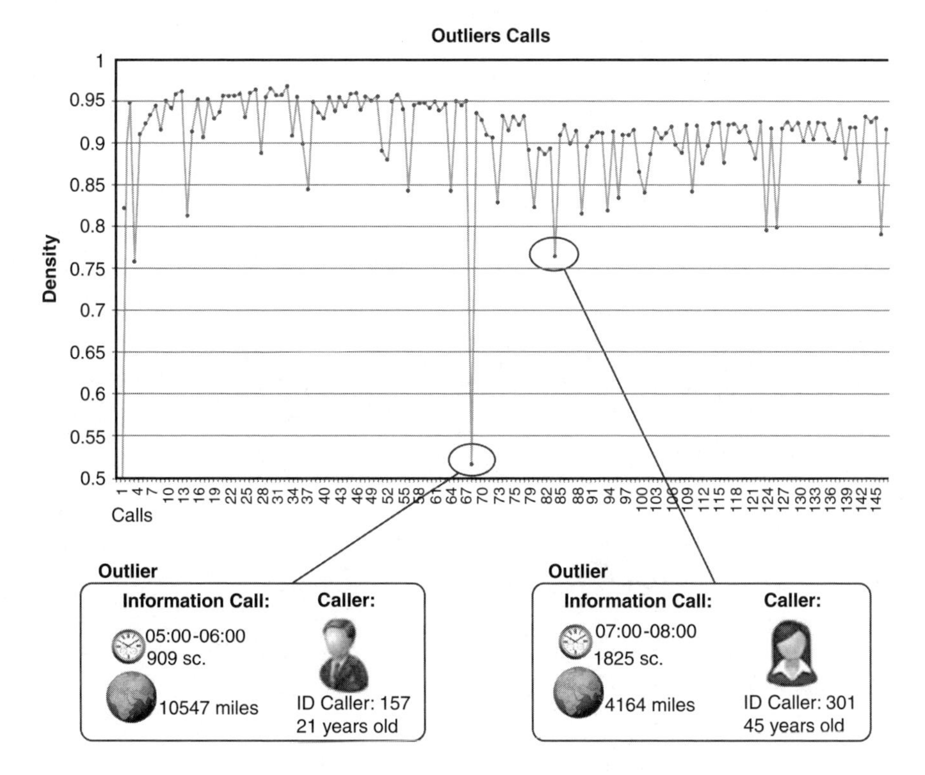

Figure 2.6 An illustration of using RDE for automatically identifying untypical/strange calls and respectively callers (adapted from STAKE project report (Angelov *et al.*, 2011))

2.7 Conclusions

In this chapter the basic notions from probability theory were introduced. The aim was not to provide a detailed description of what is a very broad topic (the reader can find more details, for example, in (Hastie, Tibshirani and Friedman, 2001; Duda *et al.*, 2000; Bishop, 2009)), but primarily to introduce concepts that will be used later on. At the same time, the author introduced his own more critical view on probability theory, which differs from the rather dogmatic view that currently dominates. Moreover, a significant innovation was also introduced regarding density estimation in the form of RDE that can be very useful on its own for novelties/anomalies detection and identification as well as in complex systems structure identification, as will be detailed later in the book.

Philosophically, the concept is considered to be rooted in *association* of a current observation (or measurement or evidence) with previously observed ones in terms of closeness in the data space (in terms of some proximity measure) rather than in terms of frequency of appearance (as is the case with the objective frequentistic approach to probability theory) or in terms of belief (as is the case with the subjective approach to probability theory and, similarly with the fuzzy logic). This allows having valid conclusions with very few observations and exact calculations. In probability theory the theoretical conclusions are ***only*** valid fully for an infinite number of observations.

3

Fundamentals of Machine Learning and Pattern Recognition

In this chapter a brief introduction to the main elements of machine learning and pattern recognition will be made that are related to the ALS such as normalisation, proximity measures, clustering, classification. They play an important role in automatic system structure identification, as will be detailed in Chapter 5, Part II.

3.1 Preprocessing

In machine learning the data is often represented as a multivariate set (in the offline case) or stream (in the online case). The number of objects/samples are characterised by more than one feature (sometimes also-called attribute in decision making, observation in data mining, measurable variable in control theory, or, simply, input). Let us denote the number of features by n:

$$x = (x_1, x_2, \dots , x_n)^T \tag{3.1}$$

In the offline mode the following matrix of observations/inputs can be formed:

$$X = \begin{bmatrix} x_{11} & x_{12} & \dots & x_{1n} \\ x_{21} & x_{22} & \dots & x_{2n} \\ \dots & \dots & \dots & \dots \\ x_{N1} & x_{N2} & \dots & x_{Nn} \end{bmatrix} \tag{3.2}$$

Autonomous Learning Systems: From Data Streams to Knowledge in Real-time, First Edition. Plamen Angelov.

where

N is the total number of observations/data samples recorded; in the online mode this will be replaced by k – the current data sample assuming that the first data sample has index *1*; each row of the matrix refers to a data sample/observation characterised by the features/attributes in the columns; an element, x_{ij} denotes the jth feature of the ith sample.

3.1.1 Normalisation and Standardisation

If the data of different columns (different features/inputs) lie within significantly different ranges (which in real problems is often the case) in order to apply correctly the distance measures one needs to normalise or standardise the data.

Normalisation is often made by forcing/mapping the data to the range [0;1]. This is done by the following formula (Duda, Hart and Stork, 2000):

$$\hat{x}_j = \frac{x_j - \underline{x}_j}{\overline{x}_j - \underline{x}_j} \qquad j = [1, n] \tag{3.3}$$

where

$\hat{x}_j$ denotes the normalised value of the jth feature/input;

$\underline{x}_j$ and $\overline{x}_j$ denote, respectively, the minimum and maximum values for the same feature/input.

This transforms the original data space into a hypercube with unit dimensions (Figure 3.1):

Obviously, this formula requires the range (and, respectively, the minimum and maximum values) per feature/input to be known and fixed, which is in contradiction with the requirement for online processing (except in some rare cases when the ranges

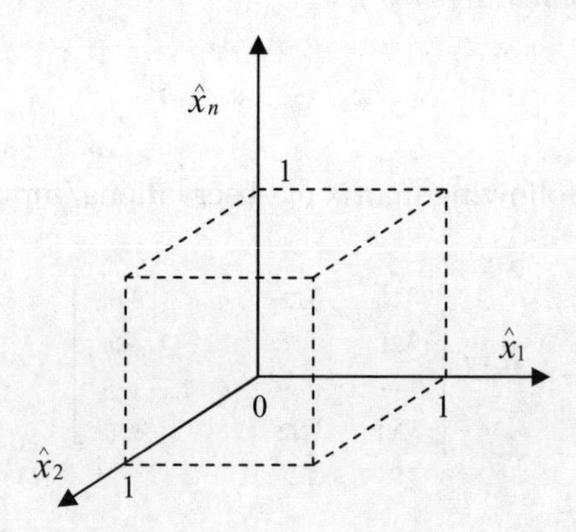

Figure 3.1 Unit hypercube in which all normalised data lie (including on the edges and vertices)

are well known in advance). It is possible to update the normalised value each time when the range changes (Zhou and Angelov, 2007), but this also affects the dynamic model update and becomes practically cumbersome.

A more suitable (from practical applications point of view) operation for conversion of the data into comparable form is the so-called standardisation (Duda, Hart and Stork, 2000):

$$\hat{x}_j = \frac{x_j - \mu_j}{\sigma_j} \qquad j = [1, n] \tag{3.4}$$

Obviously, the mean and variance are not necessarily the same for each feature/input; moreover, the distribution of the data per feature/input may not be of the same type. However, it can be proven (using Chebyshev theorem (Papoulis, 1991)) that for an arbitrary distribution the probability that the normalised data will lie outside the range [−3;3] is less than 1/9. The Chebyshev inequality can be given as follows:

$$P\left(|X - \mu| \geq k\right) \leq \frac{\sigma^2}{k^2}$$

From the Chebyshev's inequality it is obvious that the mean and the variance are enough to determine the limits of the probability within which the variable, X will lie. For example, if $k = 3\sigma$ the inequality becomes the well-known 3σ condition:

$$P\left(|X - \mu| \geq 3\sigma\right) \leq \frac{1}{9}$$

If the distribution of the data is Gaussian then the probability that the data lies outside the interval [–3;3] is even lower – around 0.3% (Duda, Hart and Stork, 2000). Obviously, one can derive similar conditions for 2σ, 6σ, and so on.

Standardisation is very convenient for online data because it is possible to update both, mean and variance online (as shown by Equations (2.31) and (2.41)).

3.1.2 Orthogonalisation of Inputs/Features – rPCA Method

3.1.2.1 The Basics of the PCA Method

When performing analysis of complex data one of the major problems stems from the number of variables involved. Analysis with a large number of variables, generally, requires a large amount of memory and computation power or an algorithm that overfits the training sample and generalises poorly to new samples. The inputs/features are often correlated and, in some problems, the number of available/measurable inputs/features is high (of the order of hundreds or more). Therefore, in such cases a dimensionality reduction is desirable. An approach that is widely used to address both of the above problems (reducing correlation between the inputs by so-called orthogonalisation and reducing complexity in terms of number of inputs used) is the so-called principle component analysis, PCA (Hastie, Tibshirani

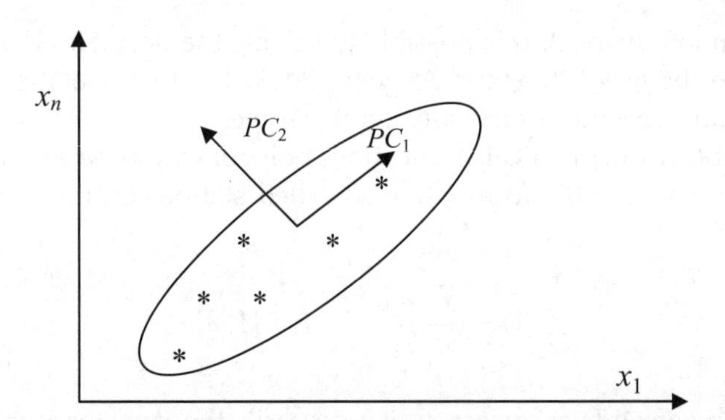

Figure 3.2 The idea of PCA method for orthogonalisation of inputs/features

and Friedman, 2001). In principle, the PCA approach is offline, but recently, online/recursive versions were also introduced (Dagher, 2010). The idea of the PCA approach is to make a transformation of the original set of inputs/feature into a new set of inputs that are, orthogonal (independent/perpendicular). This is illustrated in Figure 3.2.

The new inputs/features are formed as a result of a linear combination of the existing ones in such a way that the variation is maximised (see Figure 3.2) and the interdependence between the new input variables is nullified. The newly generated inputs/features, which are orthogonal to each other and capture most of the variance in the original data are called *principal components*, PCs. PCA transforms the data into new features space (Figure 3.2) where most of the variance is contained in the first few principal components and the remaining PCs (see axis PC_2 in Figure 3.2) can be ignored. This leads to reducing the dimensionality of the inputs used.

3.1.2.2 Offline PCA

The offline procedure is based on the so-called singular value decomposition, SVD (Duda, Hart and Stork, 2000). The multivariate matrix, X (see Equation (3.2)) can be factorised as follows:

$$X = UMV^T \tag{3.5}$$

where

matrix U is a $n \times n$ matrix of eigenvectors of XX^T, which performs a rotation of the original axes of the inputs;

M is a $n \times N$ rectangular diagonal matrix with non-negative real numbers on the diagonal which performs scaling along the rotated axes, and the $N \times N$ matrix V is the matrix of eigenvectors of X^TX, which performs another rotation.

The new coordinate system that is orthogonal and optimal (in terms of maximum variance kept) is determined by the so-called eigenvectors, u that are solutions of the equation:

$$\chi u = \Sigma u \tag{3.6}$$

where

χ denotes the eigenvalue;
Σ denotes the covariance matrix, not sum.

By selecting the largest eigenvalues only we can limit the complexity and the number of inputs actually used. As an extreme, they can be as many as the original number of inputs, n, but in practice a significantly smaller number, p is used with the largest eigenvalues containing the largest proportion of the variance of the data.

The PCA procedure starts with removing the mean value from the data (which is an element of the standardisation procedure, see Equation (3.4)):

$$\hat{x}_i = x_i - \mu \quad i = 1, 2, \ldots, N \tag{3.7}$$

Next, the scatter matrix, S is obtained via:

$$S = \sum_{i=1}^{N} \hat{x}_i \hat{x}_i^T \tag{3.8}$$

and then compute the eigenvectors, u with the largest eigenvalues, χ of the scatter matrix S.

After that, a projection matrix is made from the largest eigenvectors $U = (u_1, u_2, \ldots, u_p)$ where p is the number of eigenvectors selected. The transformed data will then be

$$\chi = U^T \hat{x}_i$$

3.1.2.3 Online (Recursive) Version of PCA, rPCA

Recently, Dagher (2010) proposed a recursive version of PCA algorithm that is suitable for our purposes. The recursive expression of the PCA starts from Equation (3.6). This is combined with the expression (Dagher, 2010):

$$v_k = \frac{1}{k} \sum_{i=1}^{k} x_i x_i^T u_i \tag{3.9}$$

where v_k is the estimate of u after the kth time step.

From the properties of the eigenvector and eigenvalues we get $\chi = \|u\|$ and $u = v/\|v\|$; u_i is then set to $v_{i-1}/\|v_{i-1}\|$. Now, the eigenvector can be estimated as follows (Dagher, 2010):

$$v_k = \frac{1}{k}\sum_{i=1}^{k} x_i x_i^T \frac{v_{i-1}}{\|v_{i-1}\|} \tag{3.10}$$

Replacing the sum, with the recursive expression we get:

$$v_k = \frac{k-1}{k} v_{k-1} + \frac{1}{k} x_k x_k^T \frac{v_{k-1}}{\|v_{k-1}\|} \tag{3.11}$$

The first component is initialised by:

$$v_1 = x_1 x_1^T \frac{v_0}{\|v_0\|};\ v_0 = x_1 \tag{3.12}$$

Each time a new data sample is provided, the eigenvectors are updated. Eigenvectors are stored in a decreasing order, starting from the largest eigenvalues. The second eigenvector is estimated by subtracting the projections of the data on the estimated first eigenvector, v_1. The new projected data x_1 on the basis of eigenvector v_2 can be computed as follows (Dagher, 2010):

$$x_2 = x_1 - x_1^T \frac{v_1}{\|v_1\|} \frac{x_1}{\|v_2\|} \tag{3.13}$$

3.2 Clustering

A well-known technique from machine learning for partitioning the data space based on the data pattern alone is *clustering* (Duda, Hart and Stork, 2000; Hastie, Tibshirani and Friedman, 2001). An important attractive feature of clustering is that it is an unsupervised learning method. Another important characteristic of this method is that it easily operates over high-dimensional data vectors.

The aim of clustering, in principle, is to 'best' separate the data into groups. However, there is not necessarily a unique separation of a certain amount of data. For example, in Figure 3.3 it is easy to see that one can identify different numbers of clusters that all seem logical.

There is a number of criteria of optimality for clustering data. The typical aim of clustering is to find natural groupings in such a way that the data points in a cluster are as similar as possible and the data points from different clusters are as dissimilar as possible (Duda, Hart and Stork, 2000):

$$J(w) = \frac{w^T S_B w}{w^T S_w w} \to \max \tag{3.14}$$

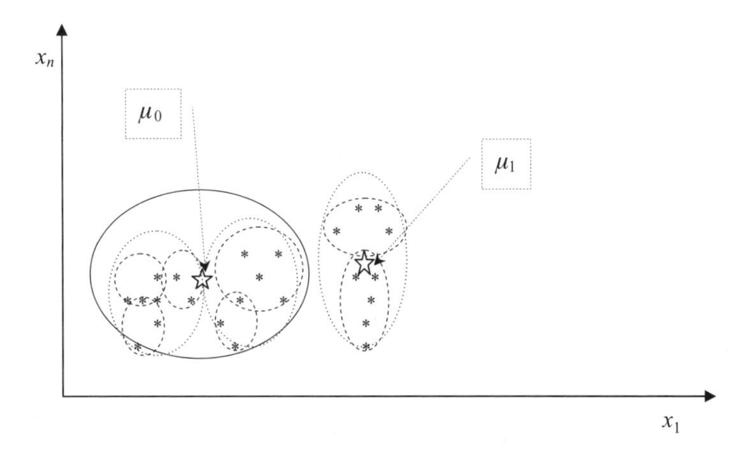

Figure 3.3 Clustering; different number of clusters (with different shape and parameters) can be built with the same data (using dotted, dashed or solid lines)

where

$S_B = (\mu_0 - \mu_1)(\mu_0 - \mu_1)^T$ is between-cluster scatter, and
$S_W = \sum_{x \in L_0} (x - \mu_0)(x - \mu_0)^T + \sum_{x \in L_1} (x - \mu_1)(x - \mu_1)^T$ is within-cluster scatter
w denotes so-called canonical variables that define the (best/optimal) line of separation; the optimal solution for w leads to $w = \Sigma^{-1}(\mu_1 - \mu_0)$

Clustering is also intrinsically related to complex systems structure identification when they are defined as multimodel systems. In this case, however, the aim of clustering is somewhat different from the aim of the clustering *per se*. In multimodel system design each cluster corresponds to a local simple model (subsystem) and, therefore, all the data samples/points that belong to a cluster are described by this local model. The overall output is usually generated as a weighted sum of local outputs and, therefore, local models are not mutually exclusive, but rather cooperative. Therefore, the data space partitioning/clustering for complex system structure identification tolerates an overlap in the clusters, unlike the conventional clustering itself.

An important problem in clustering, in general, and specifically in clustering for complex system structure identification, is the way to determine the number of clusters. Most of the clustering approaches, such as k-means (Duda, Hart and Stork, 2000), fuzzy C-means (Bezdek, 1974), Gustafson–Kessel (Gustafson and Kessel, 1978), and so on. assume that the number of clusters is prespecified. Subject to this assumption they determine the position of cluster centres and cluster parameters (radius) that minimise the objective function (3.14).

The questions 'how to determine the most appropriate number of clusters?' and 'how the number of clusters influences the result?', however, remain. Some other clustering methods, such as hierarchical clustering, *k*-nearest neighbours, learning vector quantisation (LVQ) approach and so on. (Duda, Hart and Stork, 2000) rely on thresholds, which has to be set up and influence the result.

A clustering approach that will be suitable for complex systems design must be able to:

- determine the most appropriate number of clusters from data distribution alone; and
- use a minimum amount of *prior* knowledge in the form of thresholds and algorithm parameters.

The aim of any clustering method is to find the following two items:

a. the focal points (centres) of the clusters;
b. the boundaries of the clusters.

The focal point of the cluster may not necessarily be the geometrical centre (mean) but it can be a prototype chosen to serve the role of the focal point. This is the case in prototype-based clustering approaches such as Mountain and subtractive clustering (Yager and Filev, 1993; Chiu, 1994). The boundaries of the clusters can be hypercubic, hyperspherical, hyperellipsoidal and so on.

One also needs to select the type of proximity and dissimilarity measure used in the clustering method, which will be described in more detail in the next subsection. Once the focal points of the clusters are defined any data points/samples can be assigned to the nearest cluster using the distance measure of choice (Euclidean, Mahalonobis, cosine, etc.).

Fuzzy clustering considers the more realistic case when the data points can belong to more than one cluster at the same time (with different degree of membership). This is particularly true when the semantic meaning of the linguistic terms used to define cluster boundaries is fuzzy rather than crisp (de Oliveira, 1999).

3.2.1 Proximity Measures and Clusters Shape

While clustering, various similarity measures can be considered; one of the most commonly used one is the distance between data samples. There are different types of distances such as

- Euclidean;
- Mahalonobis;
- cosine;
- Minkowski, and so on.

We will mostly use the first two. For example, Euclidean distance between two data samples (x_j and x_k) is defined as (Duda, Hart and Stork, 2000):

$$\delta_{jk}^2 = \|x_j - x_k\|^2 = (x_j - x_k)(x_j - x_k)^T = \sum_{i=1}^{n} (x_{ji} - x_{ki})^2 \tag{3.15}$$

The distance between a data sample, x_k and the cluster centre/prototype, μ_k is called *norm* (Duda, Hart and Stork, 2000):

$$\delta_k^2 = \|x_k - \mu_k\|^2 = (x_k - \mu_k)(\mu_k - x_k)^T = \sum_{i=1}^{n} (\mu_{ki} - x_{ki})^2 \tag{3.16}$$

Mahalonobis distance also takes into account the covariance:

$$\delta_{jk}^2 = (x_j - x_k)\Sigma^{-1}(x_j - x_k)^T \tag{3.17}$$

One can also define in a similar to Equation (3.16) manner the Mahalonobis norm:

$$\delta_k^2 = (x_k - \mu_k)\Sigma^{-1}(\mu_k - x_k)^T \tag{3.18}$$

Even if the data is normalised or standardised, the shape of the clusters depends on the type of the distance metric used. For example, Euclidean distance does not differentiate between the features/inputs and, therefore, the cluster shape is circular or, in general, a hyperspherical. The Mahalonobis type of distance gives more weight to the inputs/features which have higher variance and, thus, in general, the cluster shape is hyperellipsoidal (Figure 3.4).

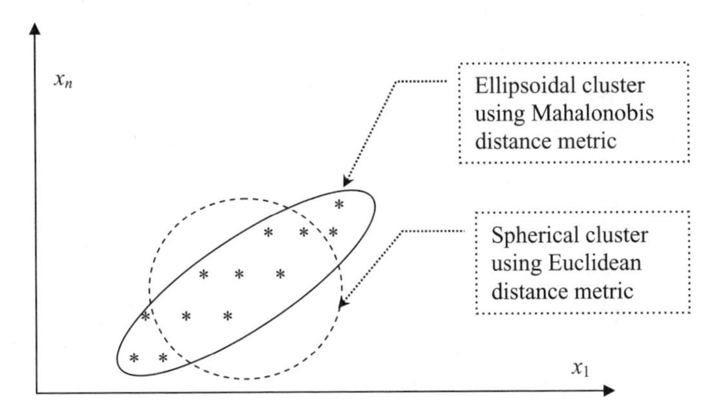

Figure 3.4 The difference in the shape of the clusters formed with the same data using different distance metrics

3.2.2 Offline Methods

Clustering methods can be broadly grouped into offline and online. As a separate group the author did introduced evolving clustering (Angelov, 2004a) in which not only data samples are considered one by one (as in online methods), but also the number of clusters is not fixed, but rather evolves dynamically.

Some important clustering methods will be briefly outlined. Other widely used offline clustering methods will not be described, but can easily be found in some of the many books on the subject such as (Duda, Hart and Stork, 2000; Bishop, 2009). These include (but are not limited to); hierarchical clustering and k means methods.

In what follows, the so-called Mountain clustering method (Yager and Filev, 1993) and its modification known as Subtractive clustering method (Chiu, 1994) will be briefly described. Both these approaches do not require the number of clusters to be predefined and extract this number form the data distribution alone. Additionally, they are prototype-based clustering techniques as opposed to the others that are mean- (centre-) based. For that reason, clusters' focal points are really existing data samples instead of virtual, abstract nonexisting (possibly nonfeasible) points as it is the case very often with the mean-based methods.

3.2.2.1 A Brief Introduction to the Mountain Clustering Method

Mountain clustering (Yager and Filev, 1993) is an important algorithm that can be used as a first step to more involved algorithms or independently aiming to generate initial cluster centres. In Mountain clustering a grid (Figure 3.5) is formed by discretising each dimension of the data space into equidistant intervals.

The intersection of the grid lines are called *nodes* and are the potential cluster centres (focal points). A *mountain* function, M is defined that is related to the density of neighbouring data points and is used to calculate the *potential* (density) of each grid point (node) to become a cluster centre.

The value of the function is high for a node with many neighbouring data points. For all the nodes the *mountain* function is calculated and the node with the highest value

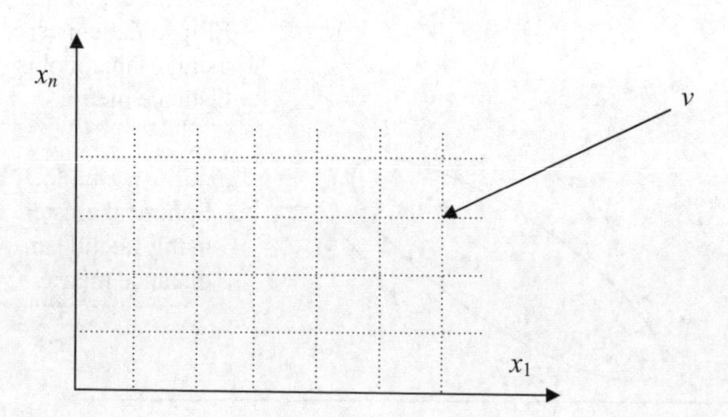

Figure 3.5 The grid of the Mountain clustering method; v denotes the node of the grid at which the *potential* is calculated

is selected as the first cluster centre. To determine the next cluster centre, an amount proportional to the distance from the point to the first cluster centre is subtracted from the current *mountain* function value of each of the nodes (Yager and Filev, 1993):

$$M_i\left(v_j\right) = \sum_{i=1}^{N} e^{-\frac{\left\|x_i - v_j\right\|^2}{2\sigma^2}} \qquad j = [1, \Gamma] \tag{3.19}$$

where

N is the total number of data points;
Γ is the number of points in the grid.

Thus, the nodes close to the first cluster centre will have a higher reduction of their value in comparison with the distant nodes. This ensures that nodes closer to the cluster centre are not selected as new cluster centres. Then, the node with the next highest value of the *mountain* function is chosen as the next cluster centre.

This process of selecting cluster centres and subsequently reducing the *mountain* function value continues until a threshold (expressed as a percentage of the first maximum) is reached. The algorithm is simple, however, computationally expensive for high-dimensional data. Each iteration requires evaluation of $O(m^n)$ nodes (considering equal number of grid lines in all dimensions) (Baruah and Angelov, 2010). The generation of number of clusters is sensitive to the grid resolution that also provides a trade-off between accuracy and computational complexity. In addition, this method requires a threshold value to be predefined as a termination criterion.

3.2.2.2 Subtractive Clustering Method Outline

Subtractive clustering is an improved version of the Mountain clustering method with the single, but significant difference that data points themselves are considered as candidate focal points (cluster centres) instead of the grid points. This method also assumes that the data points are normalised and bounded by a hypercube. For every data point a *potential* value (density) is calculated and the point with the highest potential value is selected as the first cluster centre.

The potential value is inversely proportional (reciprocal) to the sum of distances from the data point to all other data points; that is the smaller the sum distances to all other data points the higher the potential of that point (data sample) see also Figure 2.1. The neighbourhood of a data point is defined by a constant (radius, r); data points outside the neighbourhood do not have a significant influence on the potential value.

Similar to the Mountain method, the next step is to reduce the potential/density of all data points by an amount that is dependent on their distance to the cluster centre. In this way, the points closer to the cluster centre have smaller chance to be selected as the next cluster centre. The next cluster centre is the point with the remaining maximum potential, and so on. Two threshold values are defined that control the termination of the clustering process.

If the ratio of the potential (P_k) of the current data point (x_k) and the potential of the first cluster centre (P_1) is greater than an *upper threshold* value then x is accepted as the cluster centre and the process continues. If this ratio is less than a *lower threshold* value then x_k is rejected and the process terminates. If the ratio lies between the two threshold values then the smallest distance ($\delta_{\min}$) between x_k and existing clusters is determined and the following condition is examined:

$$\text{IF}\ \left(\frac{d_{\min}}{r} + \frac{P_k}{P_1} \geq 1\right)$$

THEN (x is set as the new cluster centre and the process continues)

ELSE (it is rejected and the data point with the next highest potential is selected and tested for the above conditions).

Although, the computational complexity increases linearly with the dimension of the data set, it is quadratic, not exponential. However, this algorithm requires certain critical parameters (potential-related, neighbourhood and threshold values) to be predefined. The algorithm of the Subtractive clustering algorithm is available in MATLAB® function *subclust.m*.

3.2.2.3 Gustafson–Kessel Clustering Algorithm

Gustafson–Kessel (GK) algorithm (Gustafson and Kessel, 1978) is an extension of the fuzzy c-means (FCM) algorithm (Bezdek, 1974) that itself is an extension of the k-means algorithm. The GK algorithm uses Mahalanobis norm that allows generating clusters of various size and shapes other than spherical. Each cluster is characterised by a centre and a covariance matrix that are the parameters of the cluster prototype. The use of the Mahalonobis norm allows the shape of the clusters to reflect the data better, because the eigenstructure of the cluster covariance matrix represents the shape and orientation information of the cluster. If the matrix is restricted to a diagonal form then axis-parallel clusters are generated. The disadvantage of this algorithm is that it is computationally more intensive if compared to FCM due to the involvement of matrix inverse calculations while updating the covariance matrix. Moreover, it is sensitive to initialisation of the parameters.

The algorithm starts with a random allocation of data samples to clusters (note that this approach similarly to *k*-means, FCM and most of the other clustering approaches requires the number of clusters to be prespecified). The mean values of each cluster are then calculated based on this initial random allocation and fuzzy memberships to each cluster are defined. Then, the reallocation process starts aiming minimisation of a criterion of optimality using a gradient-based method. Reallocation iterations continue until no significant improvement of the criterion is observed or no actual reallocation takes place.

3.2.2.4 Mean Shift Clustering Algorithm

The *mean shift* algorithm (Fukunaga and Hostetler, 1975; Comaniciu and Meer, 2002) is based on the KDE and the gradient-based optimisation aiming to identify the peaks of the probability density function (pdf). One limitation of this approach is that it is iterative and multipass (it starts from each of the available data samples) and is, therefore, computationally very expensive. However, because it is based on solid theoretical foundations of both the KDE and gradient-based search concepts the results it provides are very intuitive and meaningful (they often succeed to identify all or most of the local peaks of the pdf). A shortcoming of the original approach is that the choice of the kernel and distance function influences the result somewhat.

The *mean shift* method resembles expectation maximisation (EM) algorithm. *Mean shift* has a single parameter – kernel radius, h. If the data is normalised, the values of h are problem independent. The pdf is defined through a kernel in the following way (Comaniciu and Meer, 2002):

$$p(x_k) = \frac{1}{kh^n} \sum_{i=1}^{k} K\left(\frac{x_k - x_i}{h}\right) \tag{3.20}$$

where $K(.)$ denotes the kernel function.

The expression for the *mean shift* vector is derived through the gradient of the pdf as (Comaniciu and Meer, 2002):

$$m_r(x_k) = \frac{\sum_{i=1}^{k} x_k \nabla\left(\left\|\frac{x_k - x_i}{h}\right\|^2\right)}{\sum_{i=1}^{k} \nabla\left(\left\|\frac{x_k - x_i}{h}\right\|^2\right)} - x_k \tag{3.21}$$

where $\nabla(.)$ denotes the gradient of the kernel function.

The algorithm procedure of the mean shift algorithm is outlined in Appendix B1.

3.2.3 Evolving Clustering Methods

Evolving clustering methods differ from offline methods by the fact that in evolving clustering the number of clusters can evolve; that means, grow and shrink, be increased or reduced. In this respect, the first method that will be considered, incremental vector quantisation (VQ) is more limited in the sense that the number of clusters does not shrink – they can only grow.

3.2.3.1 Incremental VQ Clustering Method

In incremental online clustering methods the data samples are supposed to be used one by one (sample by sample), while in the offline methods all the data samples are available at once before the start of the procedure. The online methods are usually also noniterative and one pass that means that each data sample is used only once and is ***not*** memorised. An evolving version of one of the most simple and intuitive (and thus, popular) incremental online approaches for clustering – vector quantisation, VQ (Bharitkar and Filev, 2001) will be considered.

This algorithm starts with the first data sample assumed to be a cluster centre/ prototype. Each new data sample has two options:

a. to be assigned to an already existing cluster (if there are more than one existing clusters at a time it is assigned to the one with the nearest prototype); or
b. to initiate a new cluster.

The second option is triggered when the distance between the new data sample and any of the existing cluster prototypes/means is larger than the variance of the respective cluster. If option a) is triggered then the cluster prototype to which a new data sample is assigned is updated to take into account the new data sample using Equation (2.31). The VQ algorithm is computationally very light, but tends to create a large number of clusters many of which are formed by outliers.

3.2.3.2 Evolving Clustering Algorithm eClustering

Evolving clustering (eClustering) method (Angelov, 2004a) builds upon subtractive clustering. It can be seen as its evolving version, but is significantly different from subtractive clustering and only borrows its 'spirit'). It is also *potential* (density) based. The potential/density is calculated per data sample.

As a first step, the first sample of the data stream is established as the first cluster centre with a density set to 1. As the next data sample arrives, its potential/density is calculated using RDE.

Since the potential/density depends on the distance to ***all*** the data points, arrival of a new sample causes the potential/density of all the cluster centres to change. The potential/density of the new data sample is compared with the potentials/densities of ***all*** the existing cluster centres and one of the following actions is performed:

AThe new data sample is added as a new cluster centre if it has the highest potential/density (because of its high descriptive/representative power).

OR if the potential/density is lowest in comparison to all the existing cluster centres (to allow exploration of new areas of the data space);

B.If the new data sample has the highest potential/density and it is near to a cluster centre then it replaces the later one.

If both conditions are not satisfied then the data sample is added to the cluster with the closest cluster centre and then next data sample in the stream is considered. The process continues till all the samples in the data stream have been considered. The second part of condition A) ensures a good coverage of the whole data space. However, if data samples selected by the second part of condition A) are actually outliers, and not new operating regions no new data samples are associated with them in the future, then the algorithm takes into account also the support (number of samples per cluster) as a criterion to remove such clusters with low support (e.g. less than 1% of the total data samples at a particular instant of time).

One of the favourable characteristics of this algorithm is that it automatically handles the outliers because the potential/density of such data samples would be low due to their distance to the other data and the support of such clusters will be very low. Further, it does not require any user-defined threshold values or parameters like number of clusters that are usually required in other clustering methods.

In its initial form (Angelov, 2002, 2004a) eClustering was incapable of adapting the cluster radius and it was predefined before the start of the clustering process. However, later versions of eClustering (Angelov and Zhou, 2006), eClustering+ (Angelov, 2010) alleviate this drawback by adapting the cluster radius.

A measure called cluster *age* was also introduced (Angelov and Filev, 2005) to assess the quality of the clusters. The value of the cluster *age* is within the range $(0; k]$ and it determines whether a cluster is *young* (values close to 0) or *old* (value close to k):

$$A_k = k - \frac{1}{N_k}\sum_{i=1}^{N_k} k_i \tag{3.22}$$

where

N_k denotes the number of data samples associated with this particular cluster;

k_i denotes the time index/stamp/tag when a particular data sample has been associated with this cluster (at the moment of its assignment to the cluster this sample was obviously the current sample).

If a cluster is *young* it means that recent data samples are included in the cluster. So, new data sample with high *potential/density* value can replace *old* ones. This measure is applicable to all types of evolving clustering algorithms. The latest version called *AutoCluster* is described in Appendix B2.

3.2.3.3 Evolving Local Means (ELM) Clustering Algorithm

This is a recently developed clustering algorithm (Angelov, 2011; Baruah and Angelov, 2012) that stems from the popular mean-shift clustering algorithm. It is based on an *Epanechnikov type of kernel that has a very* convenient representation of the derivative (the mean shift) – see Equations (3.24) and (3.25).

Indeed, an Epanechnikov type of kernel is optimal in the sense of minimum variance:

$$K_E(x_k) = \begin{cases} \frac{1}{h}\|x_k - x_i\|^2 & if\ \|x_k - x_i\|^2 \leq h \\ 0 & \text{otherwise} \end{cases} \tag{3.23}$$

The main advantage of using Epanechnikov kernels is the fact that the *mean shift* expression (3.21) reduces to an update of the nearest mean if the new data sample, x_k is close (in terms of the radius, h and variance of the nearest cluster, σ) to it. The vector of the *mean shift* points precisely towards the existing mean and there is no need of iterations since there is only one new data sample at a time (therefore, ELM is noniterative and one pass!).

ELM starts from the very first data sample, same as *AutoCluster* and assumes that this is the first local mean (cluster prototype). With each new data sample it calculates the *mean shift* by updating the local/nearest mean and variance (Angelov, 2011; Baruah and Angelov, 2012) by:

$$\mu_j \leftarrow \frac{n_j\mu_j + x_k}{n_j + 1} \tag{3.24}$$

$$\sigma_j \leftarrow \frac{n_j\sigma_j + n_j\mu_j^2 x_k^2 - \left(n_j + 1 - \mu_j^2\right)}{n_j + 1} \tag{3.25}$$

where

μ_j denotes the mean of the jth cluster;
σ_j denotes the variance of the jth cluster;
n_j denotes the number of data samples associated with the jth cluster.

If the new point is further away from any of the existing local means it forms a new cluster. It is important to stress that it is optimal if the data distribution is convex. The algorithm of the ELM method is described in more detail in Appendix B3 and examples of its application to identify local peaks and to image segmentation are illustrated in Figures 3.6–3.8.

3.2.3.4 Evolving GK-like Algorithm

An online and evolving version of GK algorithm was proposed by Georgieva and Filev (2010) that was based, however, on an approximate learning of the inverse covariance matrix and determinant of the covariance matrix. An exact expression of the recursive update of the inverse covariance matrix and its determinant was proposed recently by Angelov, Kolev and Markarian (2012) based on which one can develop a GK-like evolving algorithm following the line of reasoning proposed in

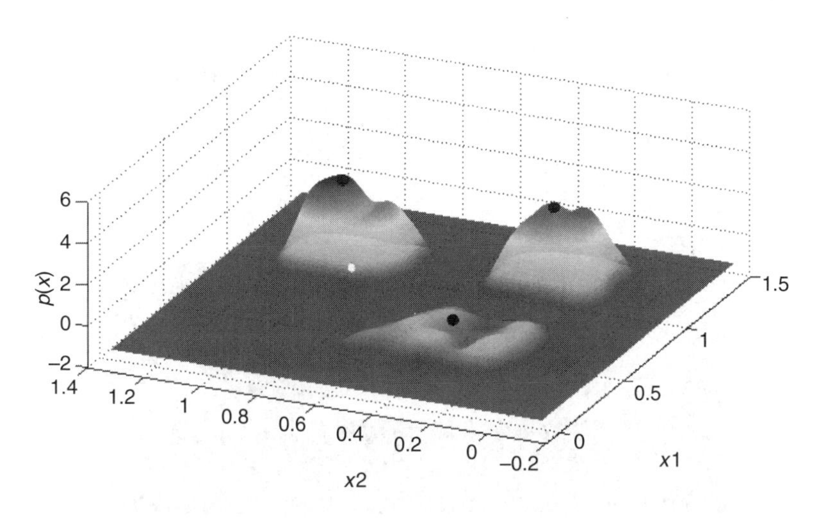

Figure 3.6 An example of applying the ELM clustering method to identify the local peaks of the density function (adapted from Baruah and Angelov, 2012)

(Georgieva and Filev, 2010). This algorithm provides clusters with ellipsoidal shape due to the use of a Mahalonobis-type distance.

In what follows, the proof of the analytical (closed from) derivation of the expressions for the inverse covariance matrix and determinant of the covariance matrix is provided (Angelov, Kolev and Markarian, 2012). If the aim is to calculate the determinant of the covariance matrix in time $k+1$, $|\Sigma_{k+1}|$. Starting from the expression of the

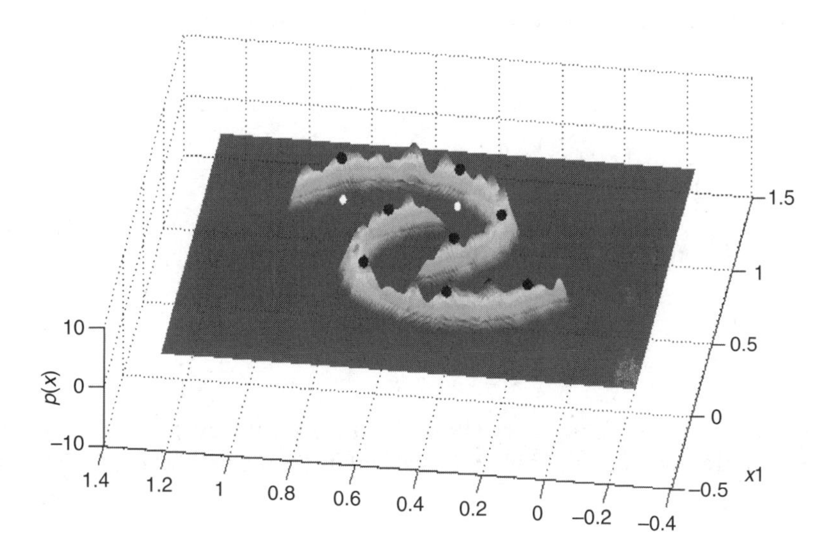

Figure 3.7 Application of the ELM clustering method to identify multiple local peaks of the density function (adapted from Baruah and Angelov, 2012)

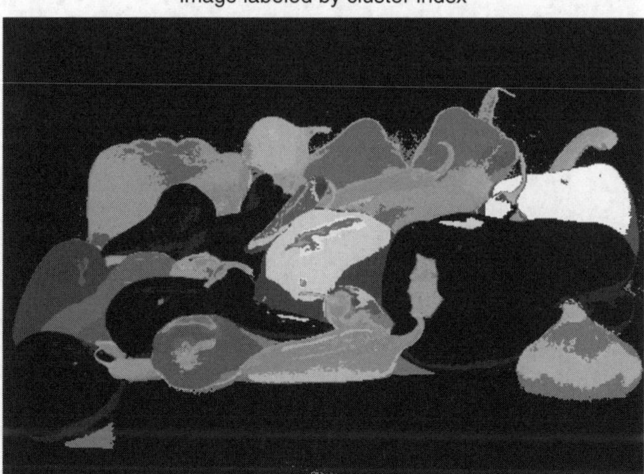

Figure 3.8 Using the ELM clustering method to segment an image (adapted from Baruah and Angelov, 2012)

covariance matrix

$$\Sigma_k = \frac{1}{k}\sum_{i=1}^{k}(x_k - \mu_k)(x_k - \mu_k)^T = \frac{1}{k}\sum_{i=1}^{k} x_i x_i^T - \mu_k \mu_k^T \tag{3.26}$$

Let's denote the first element in Equation (3.26) the mean inner product of the data item, x by itself as:

$$\Phi_k = \frac{1}{k}\sum_{i=1}^{k} x_i x_i^T \tag{3.27}$$

Then, the $(k+1)^{th}$ element can be expressed recursively as:

$$\Phi_{k+1} = \frac{k}{k+1}\Phi_k + \frac{1}{k+1}x_{k+1}x_{k+1} \tag{3.28}$$

Using the matrix inverse, Woodbury (1950), lemma (see also Appendix A2) the recursive update of the inverse of Φ can be described as (Angelov, Kolev and Markarian, 2012):

$$\Phi_{k+1}^{-1} = \frac{k+1}{k}\Phi_k^{-1} - \frac{\left(\Phi_k^{-1}x_{k+1}\right)\left(x_{k+1}^T\Phi_k^{-1}\right)}{k+1+x_{k+1}^T\Phi_k^{-1}x_{k+1}} \tag{3.29}$$

Let us reorganise Equation (3.26) as follows:

$$\Sigma_{k+1} = \Phi_{k+1} - \mu_{k+1}\,\mu_{k+1}^T = \Phi_{k+1} + (i\mu_{k+1})\left(i\mu_{k+1}^T\right) \tag{3.30}$$

where $i = \sqrt{-1}$.

The inverse covariance is then:

$$\Sigma_{k+1}^{-1} = \Phi_{k+1}^{-1} - \frac{\left(\Phi_{k+1}^{-1} i\mu_{k+1}\right)\left(i\mu_{k+1}^T \Phi_{k+1}^{-1}\right)}{1 + i\mu_{k+1}^T \Phi_{k+1}^{-1} i\mu_{k+1}} = \Phi_{k+1}^{-1} - \frac{\left(\Phi_{k+1}^{-1}\mu_{k+1}\right)\left(\mu_{k+1}^T \Phi_{k+1}^{-1}\right)}{1 - \mu_{k+1}^T \Phi_{k+1}^{-1}\mu_{k+1}} \tag{3.31}$$

Let us start with an initial estimate, Σ_o by defining the (usually used) starting conditions in a covariance estimate:

$$\Phi_0 = \alpha I \qquad \mu_0 = 0 \tag{3.32}$$

where α is a small constant.

In this way, the covariance matrix will be nonsingular from the very beginning. The determinant of Σ_{k+1} can be expressed using Equations (3.26) and (3.28) as follows:

$$|\Sigma_{k+1}| = \left|\frac{k}{k+1}\Phi_k + \frac{1}{k+1}x_{k+1}^T x_{k+1} - \mu_{k+1}^T \mu_{k+1}\right| \tag{3.33}$$

The first two components in the brackets can be transformed as follows:

$$\left|\frac{k}{k+1}\Phi_k + \frac{1}{k+1}x_{k+1}^T x_{k+1}\right| = |\Phi_k|\left|\frac{k}{k+1}I + \frac{1}{k+1}\Phi_k^{-1} x_{k+1}^T x_{k+1}\right| \tag{3.34}$$

Here, we provide without proof (the proof is in the recent patent application, (Angelov, Kolev and Markarian, 2012) the following

Lemma:

$$|\Sigma_{k+1}| = \left|\frac{k}{k+1}\Phi_k + \frac{1}{k+1}x_{k+1}^T x_{k+1} - \mu_{k+1}^T \mu_{k+1}\right| \tag{3.35}$$

Based on this we can derive the exact expression for Σ_{k+1} as follows:

$$|\Phi_{k+1}| = |\Phi_k|\left(\frac{k}{k+1}\right)^{n-1}\left(\frac{k}{k+1} + \frac{\left\langle \Phi_{k+1}^{-1}\mu_{k+1}, \mu_{k+1}\right\rangle}{k+1}\right) \tag{3.36}$$

$$|\Sigma_{k+1}| = |\Phi_{k+1}|\left(1 - \left\langle \Phi_k^{-1}, x_{k+1}^T x_{k+1}\right\rangle\right) \tag{3.37}$$

Georgieva and Filev (2010) proposed the following adaptive and recursive method for covariance matrix estimation :

$$\mu_k = (1 - \alpha)\,\mu_{k-1} + \alpha x_k \tag{3.38}$$

$$\Sigma_k = (1 - \alpha)\,\Sigma_{k-1} + \alpha\,(x_k - \mu_k)\,(x_k - \mu_k)^T \tag{3.39}$$

The noniterative formula of such estimation is given by:

$$\mu_k = \sum_{i=1}^{k} w_i x_i \qquad w_i = \alpha^{i-1}\,(1 - \alpha)^{k-i+1} \tag{3.40}$$

$$\Sigma_k = \sum_{i=1}^{k} w_i\,(x_i - \mu_i)\,(x_i - \mu_i)^T \tag{3.41}$$

Suppose that $\alpha \equiv \alpha_k = 1/k$ (and only in this case) the results for the inverse and determinant of the covariance matrix (Equations (3.31) and (3.37) are similar to the ones provided by Georgieva and Filev (2010), but even then they are not exactly the same!). The expression in (Georgieva and Filev, 2010) takes the estimate at step (k−1) and, is thus, ***an approximation*** *of the* ***exact*** *expression provided here that takes the inverse for the covariance at the same step, k.*

3.3 Classification

Classification is assigning labels to input vectors which contain features. Therefore, classification requires supervision and is offline by default (it requires training data sets). To design a classifier means to determine the mapping function $x \to L$, where $x \in R^n$ is the vector of features and $L \in R^c$ is the set of labels. Labels are, usually, integer values or can be represented as such. Very often this set contains only binary values (0 or 1). The mapping itself $L = f(x)$ can have various forms, such as (but not limited to) linear or nonlinear regression, polynomial, fuzzy rule-based, neural network type, decision trees, and so on. The classifiers can be trained by learning using some supervised learning method and training data samples (with labels). Usually, this is done offline in so-called *'batch'* mode when a set of training data samples is available for training and once the training is done the classifier can be used (Marin-Blazquez and Shen, 2002; Kovacs and Bull, 2005). There are also incremental classifiers in the sense that they can update the chosen model type and structure with each new data sample or periodically. The incremental classifiers are, however, still not evolving in the sense that the classifier structure (order of the polynomial representation, rules, neurons, etc.) are fixed.

To the best of the author's knowledge the first *evolving* (***not evolutionary***) fuzzy rule-based (and, respectively, neurofuzzy) classifier in the sense of adding or

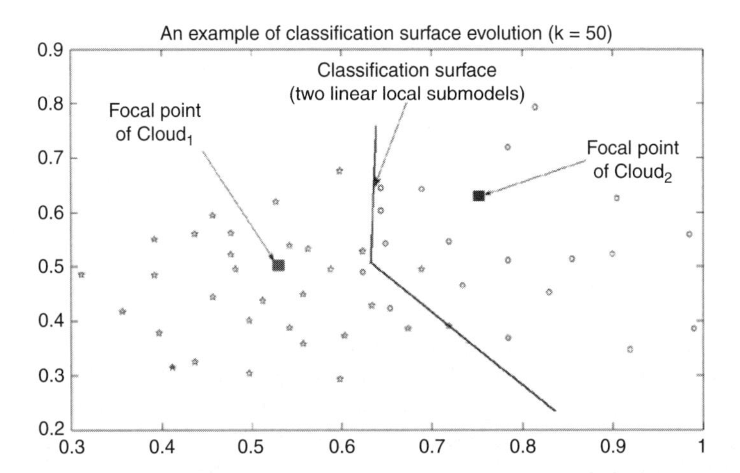

Figure 3.9a An illustration of the evolving classification surface. The solid line represents the classification surface at time step 50 based on a single-variable synthetic example case

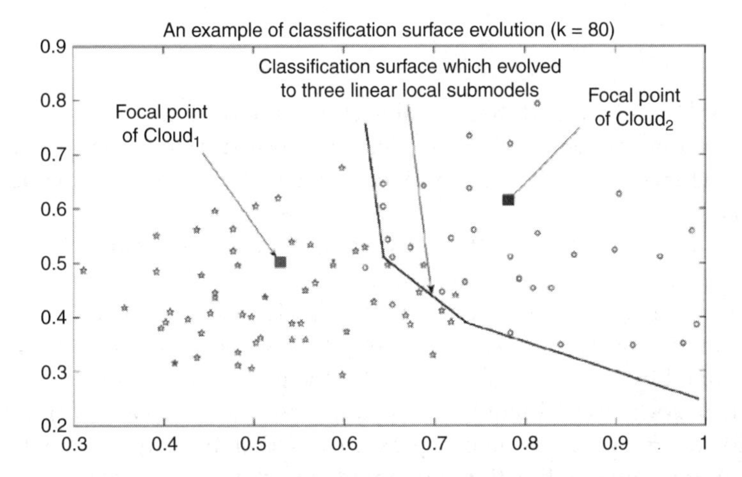

Figure 3.9b An illustration of the evolving classification surface. The solid line represents the classification surface that evolved based on 30 more data samples to three linear local submodels

removing new rules or neurons to evolve the classification surface online without complete retraining was proposed some five years ago (Angelov, Zhou and Klawonn, 2007; Angelov and Zhou, 2008)[1], see Figure 3.9. The idea is to apply

[1] In fact, there was a paper published in 2006 by the author on an evolving classifier for a specific problem (EEG signals classification) as well as a number of publications by other authors in which the term 'evolving classifier' has been used but in a completely different meaning – in the sense of evolutionary (genetic algorithms) – the difference between evolving and evolutionary is explained in detail in Section 1.6.

the estimate–update pair of operations (which is similar in adaptive control and estimation theory) to the classifiers design. The most popular methods for classification includes linear discriminant analysis, LDA (Bishop, 2009), SVM (Vapnik, 1998), neural networks, especially linear vector quantisers, LVQ. fuzzy rule-based classifiers (Kuncheva, 2000) are one of the popular classifiers, also owing to their human-intelligible linguistic form.

3.3.1 Recursive LDA, rLDA

The LDA is based on the simple principle of finding the optimal line of separation of the two (or more) classes. Obviously, this simple method cannot cope well with complex data distributions, but is a very good starting point and component in other more sophisticated methods. It was introduced in the first half of the twentieth century by Fisher (1936) and generates a linear combination of the input features:

$$L = wx^T = \sum_{i=0}^{n} w_i x_i \tag{3.42}$$

where $w \in R^{n+1}$ is a $(n + 1)$-dimensional vector of the weights.

This formula is convenient for so-called two-class classification problem ($c = 2$) while for multiclass classification problems $w \in R^{c(n+1)}$ is a $c(n + 1)$-dimensional vector and $L \in R^c$.

An underlying assumption of the LDA method is that the input features (independent variables) are *normally* distributed (mean is zero and the distribution is Gaussian, see also Appendix A1) which is often not the case in practice. LDA explicitly attempts to model the difference between the classes. The solution of the optimisation problem defined by Equation (3.14) provides the best/optimal line (or, in general, a $(n + 1)$-dimensional hyperplane) of separation between the two (or more) classes. The criterion of an input, x being in a class L is based on the projection of the multidimensional point, x onto a direction determined by w. In other words, the observation belongs to L if the corresponding x is located on a certain side of the hyperplane perpendicular to w. If we assign a threshold value, T to the expression of the line (hyperplane, in general) then the location of the plane is defined by the threshold T (it is like the offset in the equation of a line).

3.4 Conclusions

In this chapter the surface of the huge topic of machine learning and pattern recognition was barely scratched and the author does not claim to have exhaustively described it. The readers who are interested in more details on this topic are directed to more comprehensive readings such as (Duda, Hart and Stork, 2000; Hastie, Tibshirani and Friedman, 2001; Bishop, 2009, etc.).

The aim was twofold. On one hand, concepts that will be useful for the remainder of the book were defined. On the other hand, an innovative view on clustering was introduced. In terms of the innovative clustering it concerns the evolving clustering in which the number of clusters is not prespecified, but dynamically develops as a function of the density in the data space.

4

Fundamentals of Fuzzy Systems Theory

Everything is a matter of a degree.

(Australian Minister of Defence, 1908)

Fuzzy sets theory and fuzzy logic were introduced in 1965 by Lotfi A. Zadeh (Zadeh, 1965) but in a similar way as neural and evolutionary computation the theory of fuzzy sets, fuzzy logic, and fuzzy models and systems become popular only in the 1980s after the works of E. Mamdani from Imperial College, London (Mamdani and Asilian, 1975) on fuzzy controllers and T. Takagi and M. Sugeno from Japan (Takagi and Sugeno, 1985) on fuzzy modelling. It is somewhat similar to the delay between the first publications on a single perceptron in 1946 and the more wide use of neural networks in the 1980s and 1990s after the seminal works of Werbos (1974) and Rumelhart and McClelland (1986). In a very similar way, genetic algorithms (GA) pioneered by Holland (1975) were popularised only in 1990s after the much more practical book by Goldberg (1989) was published.

4.1 Fuzzy Sets

A fuzzy set is an extension of the normal set, with the main difference that an object can ***belong partially*** to the fuzzy set, instead of the binary choice that is used for the traditional (crisp, nonfuzzy) sets that limits the analysis to the following two options only:

a. to belong to a set ($x_j \in S_i$);
b. not to belong to the set ($x_j \notin S_i$).

If the membership (belonging) of the jth object to the ith set is denoted by ν the first case can be written as:

a. $\nu_{ij} = 1$.

The second case will then be:

b. $\nu_{ij} = 0$.

Instead, the use of fuzzy sets describing the membership/belonging to the cluster allows ***partial membership*** to a certain cluster:

$$0 \leq \nu_{ij} \leq 1 \tag{4.1}$$

It is usually required that the total membership to all sets for a data sample adds up to 1:

$$\sum_{i=1}^{C} \nu_{ij} = 1 \tag{4.2}$$

Fuzzy logic is a very powerful methodology of how to present information and knowledge by rules that have high generalisation and summarisation ability (Yager and Filev, 1994). One of the important characteristics of the fuzzy logic is that it can be expressed in a linguistic (natural language) form and, thus, be very intuitive. Fuzzy logic is a convenient, yet mathematically sound, tool to formalise the knowledge we extract from data.

Fuzzy logic can be blended with the traditional (classical) control (Wang, 1994), modelling (Babuska, 1998) and machine learning techniques, as will be demonstrated in the next chapters of this book. Fuzzy systems have a very important property – it was mathematically proven (Wang, 1992) that they can describe arbitrarily well any nonlinear (complex) function; that is, fuzzy systems are universal approximators.

A fuzzy set is described by its membership function. There exist a variety of types of membership functions, but the most commonly used ones are (Dubois, Prade and Lang, 1990; Yager and Filev, 1994):

i. triangular;
ii. trapezoidal;
iii. Gaussian;
iv. Cauchy;
v. sigmoid, and so on.

If we take the Gaussian membership function, for example, there is a superfluous similarity between membership to a fuzzy set and probability density function. Analysing

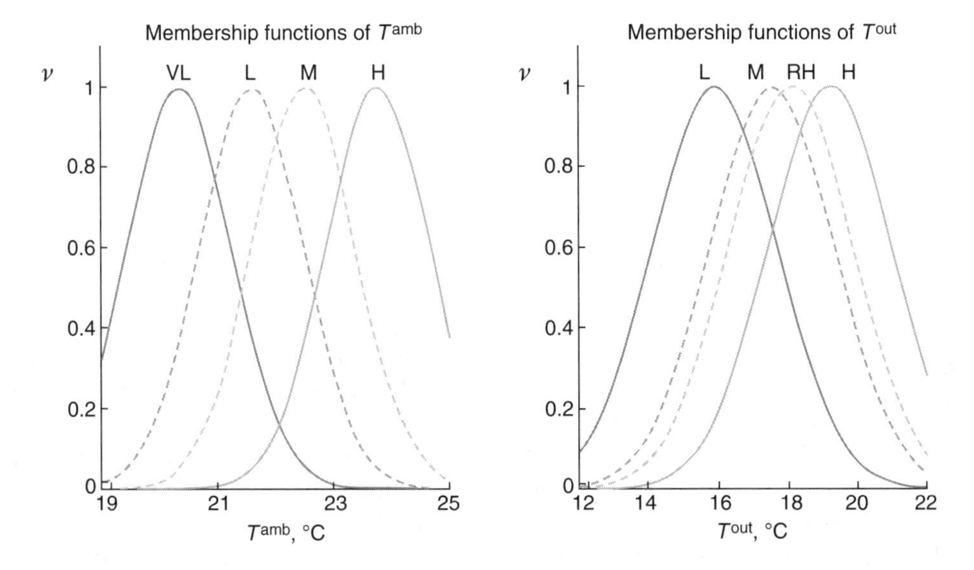

Figure 4.1 Examples of bell-shaped membership functions of ambient temperature (left plot) and outlet temperature of the air-conditioning systems heat exchanger (Angelov, 2002) which represent the linguistic variables *Very Low, Low, Medium, High* and *Relative High*

deeper, however, there are substantial differences in the meaning. In addition, formally, the membership functions to a fuzzy set have normally a maximum of 1 and the integral under the membership function is (significantly) larger than 1, unlike pdf that integrates to 1 and has maximum lower than 1. This follows from Equations (2.3) and (2.4) and, respectively, Equations (2.22) and (2.23) that are part of the Kolmogorov conditions for probability, but do not apply to membership functions of fuzzy sets and, equally, do not apply to the density function formulated by Equations (2.27) and (2.29) and used in this book.

The difference in the meaning is that the probability defines uncertainty of a fact that may take place or not, but if it takes place it is entirely available, not partially while the fuzzy sets represent uncertainty in terms of a partiality. For example, looking at the membership functions represented in Figure 4.1 if the ambient temperature is 23 °C that means that it is *partially* (with some *degree*) *Medium*, but also to a certain *degree* it will be a *High* temperature. There is some duality, some uncertainty still left even after the real measurement is made. It is not of a random nature, but of ambiguity, lack of sharpness type. One should consider the linguistic labels with some flexibility because in this specific case they relate to the air-conditioning system (otherwise a temperature of 21 °C is hardly a *Low* temperature). If we consider a similarly looking pdf (they will have, of course, maxima lower than 1) the meaning is completely different. The probability, for example, that the ambient temperature will be tomorrow 23 °C means that when tomorrow comes the ambient temperature **may** be 23 °C **or may not**. And out of N cases in N_{23} it will be. But if that is one of these cases it will not be partially 23 °C.

A more obvious example is with the pregnancy. The probability that a woman in her 20s was pregnant can be, for example, 60%, which means that out of 1 million women in their 20s 600 000 were pregnant, but (this is important to note) every single one of them was fully (not partially!) pregnant while a fuzzy set for this example is hard to formulate.

However, if we take smoking, the probabilistic example may consider the fact that, for example 12% of the people in an area are smokers, but if we take a single one (s)he will be either smoker or nonsmoker. For this particular example a fuzzy set can be defined, but it will have an entirely different meaning. Namely, we can consider the degree of smoking habit and we can formulate fuzzy sets similar to the ones represented graphically with their membership function in Figure 4.1.

Then we can distinguish between '*Heavy*' smokers (for example on the horizontal axis we can have the number of cigarettes being smoked per week and then this can mean over 100), 'Occasional' (who smoke less than 10 cigarettes), and so on.

The difference in the meaning becomes clear from these practical everyday-life examples. It is also obvious that both types of uncertainties representation have their own place. Fuzzy sets represent partial truth, duality (the ability to be represented by more than one label, description, model, structure at the same time, to a certain degree. The probability represents the frequency or belief in a fact taking place (but when it takes place it is fully and entirely existing, not partially).

4.2 Fuzzy Systems, Fuzzy Rules

Based on fuzzy sets and fuzzy variables one can formulate fuzzy rules that are linguistic statements of the following type:

$$\begin{aligned} &Rule_i\text{: } IF\ (antecedent) \\ &\qquad THEN\ (consequent) \quad i = [1, R] \end{aligned} \tag{4.3}$$

where R is the number of fuzzy rules; *antecedent* and are different for different types of fuzzy rules (can be linguistic or functional/mathematical expressions).

There are several types of fuzzy rules, but two of them are widely used now:

i. so-called Mamdani or Zadeh–Mamdani type, and
ii. so-called Takagi–Sugeno (TS) called sometimes Takagi–Sugeno–Kang type.

Both are named after the researcher(s) who introduced them. In this book we will also consider a third alternative type that was recently introduced by Angelov and Yager (2010, 2012) and is called AnYa.

All three types of fuzzy rules differ by the form of their antecedent and/or consequent part. Their mechanism of producing the overall output as a fuzzy blend of local outputs that is called defuzzification is the same for all the three types.

Based on fuzzy rules one can compose a fuzzy rule-based system (classifier, predictor, controller, filter, estimator). Such systems have been applied to a range of control systems, decision making, pattern recognition and system modelling disciplines. They play a crucial role in a range of industrial applications such as consumer products, robotics, manufacturing, process control, medical imaging, financial trading and so on (Yager and Filev, 1994).

4.2.1 Fuzzy Systems of Zadeh–Mamdani Type

Sets of such linguistic fuzzy rules are known as Mamdani-type fuzzy systems and were first introduced in early 1970s by Lotfi Zadeh (Zadeh, 1975) and Abe Mamdani and his students for linguistic description of a feedback controller (Mamdani and Asilian, 1975). The (Zadeh–) Mamdani type of fuzzy rule-based systems is a collection of fuzzy rules of the form:

$$\begin{aligned} Rule_1\text{: } & IF\ (Car\_Weight\ is\ \mathbf{High})\ AND \\ & (Volume\_Cylinders\ is\ \mathbf{High}) \cdots\ AND\ \cdots \\ & THEN\ (Fuel\_Efficiency\ is\ \mathbf{Low}) \end{aligned} \tag{4.4}$$

where

High, Low, and so on are linguistic terms represented by fuzzy sets defined by their membership functions;

Fuel_Efficiency denotes the miles one can drive by a gallon – so-called miles per gallon (mpg).

In (Zadeh–) Mamdani-type fuzzy rule-based systems both the antecedent and consequent parts of the fuzzy rules are defined by fuzzy sets. In addition, the antecedent part is defined per input variable (not in a vector form) and the fuzzy sets per variable are then connected by conjunction operators (t-norm) that can be interpreted as a logical AND (Dubois, Prade and Lang, 1990).

It is well known that the linguistic information is imprecise in its nature (Kacprzyk and Zadeh, 1999) and, therefore, the formalism in the design of (Zadeh–) Mamdani-type linguistic fuzzy systems is focused on the definition of membership functions of the respective fuzzy sets describing the linguistic variables. The level of overlap between neighbouring membership functions plays a pivotal role in the flexibility and the power of the fuzzy inference. It is determined by the parameters of the fuzzy sets and affects the interpretability in the sense that the lower the overlap the more clear is the interpretability.

4.2.1.1 Linguistic Terms and Variables

The linguistic expressions are used to represent values that are imprecise such as *'young'*, *'warm'*, *'small'* and so on. A representation is called *complete* if for any value of the variable (e.g. *Car_Weight*) there is at least one fuzzy set that covers and describes it. An example of a complete representation is the Gaussian membership function, while examples of noncomplete representations are, for example, the triangular and trapezoidal membership functions. Having the ability to represent the models in a linguistic form is one of the main advantages of fuzzy rule-based models. This is true for both the ability to 'extract' knowledge from data (streams) and for encoding and formalising existing expert knowledge and integrating it with data-driven models. Interpretability of fuzzy rule-based models is an important aspect of their design (de Oliveira, 1999). Another important advantage of the fuzzy sets is their ability to represent partial truth, to belong to more than one class or cluster at the same time (partially). For example, it is very obvious that in life the preferences for a film, music or politician are more naturally to be represented by a fuzzy membership. Similarly, a situation or observation can possibly be described by more than one (simple) model partially.

4.2.1.2 Inference and Defuzzification

The defuzzification is usually done by one of the two widely used techniques, namely:

a. the mean of the maximum (MoM) also known as 'the winner takes all':

$$y = y_{i^*}; \quad i^* = \arg\max_{i=1}^{R} (\lambda_i) \tag{4.5}$$

b. the centroid or centre of gravity (CoG) method:

$$y = \sum_{i=1}^{R} \lambda_i y_i \tag{4.6}$$

4.2.2 Takagi–Sugeno Fuzzy Systems

Takagi–Sugeno (TS) type of fuzzy systems (Takagi and Sugeno, 1985) are currently one of the most popular types of fuzzy rule-based systems. This is mainly due to their dual nature – they combine a linguistic, fuzzy IF part with a functional (usually, linear) consequents part, for example:

Rule$_1$: IF (*Car_Weight is* **High**) *AND* (*Volume_Cylinders is* **High**)
THEN (*Fuel_Efficiency* $= a + b^*$*Car_Weight* $+ c^*$*Volume_Cylinders*)

4.2.2.1 Architecture of Takagi–Sugeno Fuzzy Systems

The overall TS-type fuzzy model can be described in the following form:

$$\begin{aligned} R_i:\ & IF\,(x_1 \sim x_{1i}^*)\,AND \cdots AND(x_n \sim x_{ni}^*) \\ & THEN\,(y_i = LM_i) \end{aligned} \tag{4.7}$$

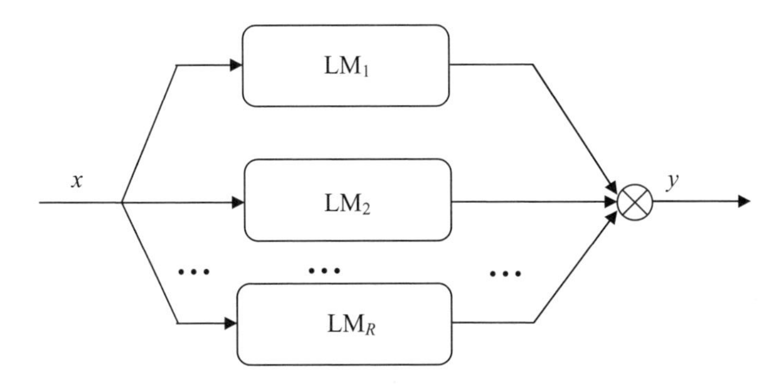

Figure 4.2 TS-type fuzzy system

where

R_i denotes the ith fuzzy rule; $i = [1, R]$;
R is the number of fuzzy rules;
$(x_j \sim x_{ij}^*)$ denotes the jth fuzzy set of the ith fuzzy rule;
$j = [1, n]$; x_i^* is the focal point of the ith rule antecedent part;
LM_i denotes i^{th} local model, $i = [1, R]$;
y is the output.

The antecedent part is a linguistic representation of the partition of the measurable input variables space into fuzzily overlapping regions that define locally valid (often, but not necessarily) linear systems.

The linguistic antecedent part of the TS fuzzy systems makes them attractive for human operators (if we compare to neural networks (NN), support vector machines (SVM) or polynomial models, for example). Their architecture, see Figure 4.2, is composed of fuzzily weighted local (in terms of data space) output linear models that can be represented in a vector form as:

$$LM_i : y_i = X^T A \tag{4.8}$$

where

$X = [1, x_1, x_2, \ldots, x_n]^T$ denotes the $(n + 1) \times 1$ extended vector of measurable variables;

$A_i = [a_{0i}\ a_{1i} \cdots a_{ni}]^T$ denotes the matrix of consequent parameters.

All of the R linear models describe the process in a local area defined by fuzzy rules and are blended in a fuzzy way (see Equation (4.6)) to produce the overall output, y which is nonlinear in terms of the measurable/observable input variables, x; but is linear in terms of the parameters, A. The global (in terms of data space) model can be described in a vector form as:

$$y = \psi^T A \tag{4.9}$$

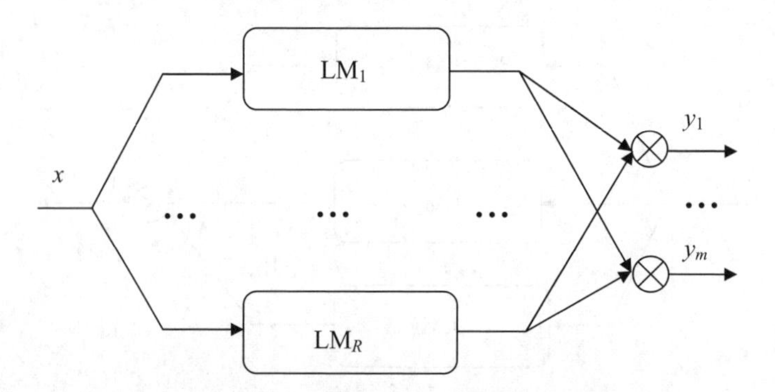

Figure 4.3 A MIMO TS-type fuzzy system

where

$\psi = [\lambda_1 X^T, \lambda_2 X^T, \ldots, \lambda_R X^T]^T$ is a vector of the measurable variables that are weighted by the normalised firing strength (or activation levels) of the rules,

λ_i, $i = [1, R]$ is the normalised firing level of ith fuzzy rule that is a function of x, that is $\lambda_i(x)$.

The degree of membership of certain data sample, x to any of the fuzzy rules is usually described by a Gaussian function centred at its focal point, x_i^*:

$$\nu_i = e^{-\frac{\|x - x_i^*\|^2}{2\sigma_i^2}} \tag{4.10}$$

4.2.2.2 Multi-Input–Multi-Output (MIMO) Takagi–Sugeno Fuzzy Systems

In their standard form, TS FRB were defined for the case of so-called MISO (multiple-input–single-output) models (Takagi and Sugeno, 1985), but in 2004 a MIMO (Multiple Inputs Multiple Outputs) form was introduced for the case of evolving TS models (Angelov *et al.*, 2004b), see Figure 4.3. In this formulation the output, $y_i = [y_{1i},\ y_{2i}, \ldots,\ y_{mi}]^T$ is also a vector with dimension $(m \times 1)$ that can be interpreted linguistically as:

$$\begin{aligned} R_i : IF\ (x_{1i} \sim X_{1i})\ AND \cdots AND\,(x_{ni} \sim X_{ni}) \\ THEN\ (y_{1i}\, LM_{1i})\ AND \cdots AND\,(y_{mi} = LM_{mi}) \end{aligned} \tag{4.11}$$

The dimensionality of the parameters of the local submodels of the consequent part increases and they are represented by a matrix (Angelov *et al.*, 2004b):

$$A_i = \begin{bmatrix} a_{01i} & a_{02i} & \cdots & a_{0mi} \\ a_{11i} & a_{12i} & \cdots & a_{1mi} \\ \vdots & \vdots & \cdots & \vdots \\ a_{n1i} & a_{n2i} & \cdots & a_{nmi} \end{bmatrix} \tag{4.12}$$

Thus, each element $(y_{1i}\ y_{2i} \cdots y_{mi})$ of the m-dimensional output, y_i can be represented as a separate line:

$$y_{1i} = a_{01i} + x_1 a_{11i} + \cdots + x_n a_{n1i};$$

$$y_{2i} = a_{02i} + \cdots + x_n a_{n2i}$$

and so on.

4.2.2.3 Analysis of the Inference in Takagi–Sugeno Fuzzy Systems

In the case of classical/traditional sets ('normal', two-valued, binary) logic IF–THEN rules are easy to interpret as follows:

IF (the premise/antecedent is true)
THEN (the consequent is also true/holds)

Classical rules were used in 1970s and 1980s in so-called *expert systems* (Giarratano and Riley, 1998). In the case of fuzzy (multivalued) logic these conditions are relaxed by allowing a degree of satisfaction, degree of truth, partial truth. This is closer to the complexity of the real-life situations and gives a more realistic than just 'black and white', 'true or false' picture.

Interpreting IF–THEN rules is a three step process (Yager and Filev, 1994; Dubois, Prade and Lang, 1990):

1. Perception: That means mapping all the inputs in the antecedent part to a degree of membership (lying in the interval [0;1]) to respective (linguistic) fuzzy sets.
2. Aggregate the multiple part antecedents: If there are multiple parts of the antecedent, apply a conjunction (intersection, AND) operator to find the overall degree of truth of the antecedent part. Note, that this stage is not necessary in the AnYa type of FRB.
3. Apply inference (defuzzification) method: use the degree of support for the entire rule to shape the output of the fuzzy rule-based system. Because the rule, is, generally, only partially (by λ_i) true, therefore the output is a truncated version of the fuzzy set of the output (in the case of a Mamdani-type fuzzy rule) or only a fraction (represented by a weight between 0 and 1) of the output y that the fuzzy rule produces (in the TS and AnYa type of fuzzy rule).

4.3 Fuzzy Systems with Nonparametric Antecedents (AnYa)

The new type of fuzzy rule-based systems called AnYa (Angelov and Yager, 2010, 2012) was proposed as an attempt to revise and simplify the antecedent part of fuzzy rule-based systems that for both traditional types (Mamdani and TS) is the same. At the same time, the antecedent part plays a key role in learning (as will be demonstrated in the next two chapters). It is traditionally either predetermined by regular partitioning or clustering (Babuska, 1998; Carse, Fogarty and Munro, 1996) or is optimised by supervised learning, for example computationally expensive error back-propagation (Jang, 1993) or genetic algorithms (Cordon *et al.*, 2004; Angelov and Buswell, 2003).

AnYa includes a nonparametric antecedent part of a new type which also simplifies the linguistic expression removing the need for logical AND, and the ambiguity related to the choice of the t-norm operator. In addition, the neural networks interpretation (which in the case of TS model is of a five-layer network) is simpler and reduces to a four-layer structure as will be shown later.

4.3.1 Architecture

Let us consider a complex, generally nonlinear, nonstationary, nondeterministic system that can only be described and observed by its input and output vectors, $x = [x_1, x_2, \ldots, x_n]^T$ and $y_i = [y_{1i}, y_{2i}, \ldots, y_{mi}]$, respectively. The aim is to describe the input–output dependence based on a history of observation of input–output pairs, $z_j = [x_j^T; y_j^T]^T, j = 1, 2, \ldots, k{-}1$ and current, k inputs, x_k^T only. The dimension of the vector of input–output data z_j is $(n + m)$: n dimensions of the inputs and m dimensions of the outputs.

The new alterative type of fuzzy rule-based systems (Angelov and Yager, 2010, 2012) is based on data *clouds* that are very much like clusters, but differ in several aspects. They are nonparametric. They do not use their mean or variance and, thus, they do not have a specific shape and boundary. They are just collections of data samples (points in the multidimensional data space), see Figure 4.4.

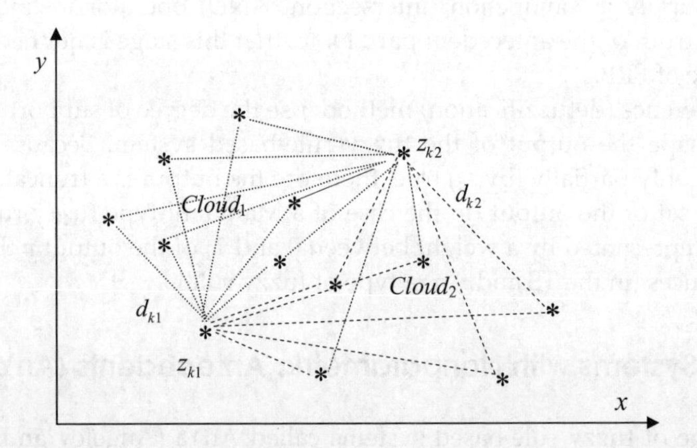

Figure 4.4 Data clouds. Data items (points) from $Cloud_1$ are linked with the data item, z_{k1} using dotted lines and, respectively, the data items (points) from $Cloud_2$ are linked with the data items, z_{k2} using dashed lines. It is obvious that data clouds (unlike clusters) have no specific shape

The main difference is that clustering approaches simplify the real data representation replacing it with a set of cluster centres and cluster zones of influence have a specific shape (hyperspherical shape if using Euclidean type distance and hyperellipsoidal shape if using a Mahalonobis-type distance). Data clouds, on the contrary, take full account of the density based on ***all*** data samples/points.

As a result, an AnYa type of fuzzy rule-based systems does not consider and does not require membership functions or fuzzy sets per scalar variable to be formulated (Angelov and Yager, 2012). In this sense, the proposed simplified fuzzy rule-based structure can also be seen as ***type 0 fuzzy sets*** (by analogy to the ***type II fuzzy sets*** (Karnik, Mendel and Liang, 1999) for which the membership functions are defined by a fuzzy set for each point of the membership function. "*Type 0*" means that there is no need to define even the membership functions per variable but it only suffice to define (calculate recursively) the *local* density.

AnYa-type fuzzy rule-based systems **do not require** an explicit definition of the membership function or even a prior assumption of its form). On the contrary, if necessary, the membership function per input variable/feature can be defined based on the ***real*** data distribution (Angelov and Yager, 2010). A very interesting and strong aspect of the proposed method is the **nonparametric** form of the data *clouds* as local building blocks of an overall complex system.

Data *clouds* are sets of previous data samples with common properties (closeness in the data space). They **directly** represent **all previous data samples.** Contrast this to the traditional membership functions that usually do **not** represent the true data distributions; instead, membership functions represent some desirable/expected/estimated (often subjectively) preferences.

The fuzziness in AnYa FRB is primarily related to the inference and the fuzzy membership of a particular data sample to (multiple or) all *clouds* simultaneously with different degree, $\lambda \in [0;1]$ defined by the local density, d (2.29). A *cloud* is described by the set of data samples that belong to it as well as by a statement of the following form:

$$(x \sim \Xi_i) \quad \Xi_i \in R^n \quad i = [1, R] \tag{4.13}$$

where

$\Xi_i \quad \Xi \in R^n \quad i = [1, R]$ denotes a *cloud;*
'$\sim$' can be read as '*is like*'.

The degree of membership to a *cloud* is measured by the normalised (using, the centre of gravity principle) *local* density for a particular data sample, x_k:

$$\lambda_{ki} = \frac{d_{ki}}{\sum_{j=1}^{R} d_{kj}} \quad i = [1, R] \tag{4.14}$$

where d_{ik} is the *local* density of the ith *cloud* for a particular data sample, x_k (index k will be further omitted for simplicity for both d and λ); it is defined by Equation (2.29); or recursively as in Equation (2.30) for the global density.

Because the fuzzy membership to a *cloud*, λ_i is normalised, they sum up to 1:

$$\sum_{i=1}^{R} \lambda_i = 1 \tag{4.15}$$

An AnYa-type FRB has the following linguistic expression:

$$\begin{aligned} &Rule_i\text{: } IF\ (x \sim \Xi_i) \\ &THEN\ (y_i = X^T a_i) \end{aligned} \tag{4.16}$$

where the degree of fulfilment of the antecedent part is determined by the normalised local density, λ_i.

4.3.2 Analysis of AnYa

Comparing the two traditional types of fuzzy rule-based systems (see Table 4.1) one can observe their similarity in terms of the antecedent (premise) part.

While both the consequent part and the defuzzification inference differ, the antecedent parts of both Mamdani and TS are exactly the same. Yet, this type of antecedent part formulation is often a stumbling block in the practical design of fuzzy rule-based systems. This is true both in the case when their design relies on real data as well as when it relies on expert knowledge. The reason is that defining membership functions per scalar variable and parameterisation of all of them requires a very high level of approximation (because the real data distributions and real problems are often not smooth and easy to describe 'per variable').

An AnYa-type fuzzy rule-based has some parallels and similarities with the Bayesian probabilistic systems in terms of their inference. Indeed, we can summarise

Table 4.1 A comparison of the three types of fuzzy rule-based systems

	Antecedent/IF Part	Consequent/ THEN Part	Defuzzification
Mamdani	Scalar, parameterised fuzzy sets	Scalar, parameterised fuzzy sets	Centre of gravity
TS	"	Functional (usually linear)	Fuzzily weighted sum (average)
AnYa	All data; nonparametric data *clouds*	"	"

Table 4.2 Duality between the inference in Bayesian (probabilistic) and AnYa-type fuzzy rule-based models

Bayesian (probabilistic)	AnYa-type FRB/NFS
$p(y \mid x)$	y
$p(y)$	y_i
$p(x \mid y)$	d_i
$p(x)$	$\sum_i d_i$
$p(y \mid x) = \dfrac{p(x \mid y)p(y)}{p(x)}$	$y = \sum_{i=1}^{R} \frac{d_i}{\sum_{j=1}^{R} d_j} y_i$

the inference of AnYa type fuzzy rule-based systems by Equations (4.9) and (4.14). Indeed the overall output, y of the AnYa-type FRB is conditioned on the input, x:

$$p(y \mid x) \rightarrow y \tag{4.17}$$

where '→' denotes 'corresponds to'.

The unconditional probability of the output can be related to the local/partial outputs:

$$p(y) \rightarrow y_i \quad i = [1, R] \tag{4.18}$$

The probability of the input conditioned on the output, $p(x \mid y)$ is related to the density of the data sample, d while the unconditional probability of the input, $p(x)$ is obviously a sum/integral of local densities. Combining, we can derive Table 4.2 that represents the duality between the AnYa type of FRB and Bayesian inference.

4.4 FRB (Offline) Classifiers

Fuzzy rule-based (FRB) classifiers consist of fuzzy rules of the following form:

$$\begin{aligned} &Rule_i\text{: } IF\ (x_1 \sim x_{1i}^*)\ AND\ \cdots\ (x_n \sim x_{ni}^*) \\ &THEN\ (L_i) \end{aligned} \tag{4.19}$$

where

$i = [1, R]$; R is the number of fuzzy rules;

$(x_j \sim x_{ji}^*)$ denotes the jth fuzzy set of the ith fuzzy rule; $j = [1, n]$;

x_{i1}^* is the focal point of the ith rule antecedent (note that this is a prototype – a real, existing data sample not the mean).

L_i is the label of the class of the ith prototype (focal point);

'∼' denotes the linguistic expression '*is like*' represented by a fuzzy set.

This type of FRB classifier is of so-called 0 order because the consequent part consist of singletons. It is also possible to have *first*-order classifiers where the consequent part consist of lines (Angelov *et al.*, 2007). Then, the value of the consequent is used for comparison with a threshold (the threshold is 0.5 if we use normalised values in the interval [0, 1]).

$$
\begin{aligned}
&Rule_i\text{: } IF \left(x_1 \sim x_{1i}^*\right)\ AND \cdots \left(x_n \sim x_{ni}^*\right) \\
&THEN \left(y_i = w_{0i} + \sum_{j=1}^{n} w_{ji} x_j\right)
\end{aligned}
\tag{4.20}
$$

For a given set of feature values, x, first, the membership functions of the fuzzy sets are determined as values between [0; 1]. The membership functions that describe the degree of association with a specific prototype can, for example, be of Gaussian form that is characterised by good generalisation capabilities and coverage of the whole feature space. Then, these values are aggregated for all features, $j = 1, 2, \ldots, n$ and the so-called firing level of (degree of confidence in) the ith fuzzy rule, τ_i is determined using an operator (minimum or product) for logical AND. Once the firing levels, τ_i per rule are determined the label can be inferred by the so-called 'winner takes all' principle:

$$
L = L_{i^*}; \quad i^* = \arg \max_{i=1}^{R}(\lambda_i) \tag{4.21}
$$

The firing level, τ of a fuzzy rule is determined by a *t-norm*, which can be represented as an inner product (Yager and Filev, 1994):

$$
\tau_i = \prod_{j=1}^{n} \nu_{ij}(x_j) \tag{4.22}
$$

or minimisation

$$
\tau^i = \min \left(\nu_{ij}\left(x_j\right) \right)_{j=1}^{n} \tag{4.23}
$$

where n is the number of fuzzy sets/linguistic terms such as *'Small'*, *'Medium'*, and so on, which are represented by their membership functions, ν.

The firing level is then normalised so that it sums up to one:

$$
\lambda_i = \frac{\tau_i}{\sum_{j=1}^{R} \tau_j} \tag{4.24}
$$

For first-order FRB classifiers the overall output is determined as a weighted sum using so-called 'centre of gravity' (CoG) principle where the weights represent the normalised firing level of a particular rule and, respectively, its local (linear) output:

$$y = \sum_{i=1}^{R} \lambda_i \overline{y}_i \tag{4.25}$$

where $\overline{y}_i = \frac{y_i}{\sum_{j=1}^{R} y_j}$ is the normalised output.

Since the normalisation leads the output, y_i to be in the region [0;1] therefore the two classes can be separated, for example, as:

$$\begin{aligned} &IF\ (y > 0.5) \\ &THEN\ (Class\ 0) \\ &ELSE\ (Class\ 1) \end{aligned} \tag{4.26}$$

The FRB classifiers can be trained/learned using training data or, alternatively, the rules can be provided by expert knowledge. Learning can be performed offline (training data are provided as a batch set) or online (training data arrive one by one in real time) or can be developed in an evolving manner (the number of rules and fuzzy sets can be dynamically changing (growing or being reduced) instead of being fixed (Angelov and Zhou, 2008).

4.5 Neurofuzzy Systems

The combination of fuzzy rule-based systems with neural networks led in 1990s to so-called neurofuzzy systems (NFS) that combine the advantages of fuzzy rule-based systems having linguistic expression, allowing partially valid simpler local models with the layered structure of neural networks that makes them convenient for learning. One particular type of neural networks – radial-basis function (RBF) type NN has also very close links with FRB because neurons in the hidden layer of an RBF (Figure 4.5) can be interpreted as fuzzy sets with Gaussian membership function.

RBF-type neural networks have three layers, namely input layer that merely passes the inputs to all neurons of the next layer, called the hidden layer; hidden layer itself that determines the closeness of particular inputs to prototypes that are determined as a result of learning; and, finally, the output layer that performs a linear combination of the (partial) outputs produced by each neuron from the hidden layer.

4.5.1 Neurofuzzy System Architecture

The architecture of a NFS is formed of several layers with specific role. Both, antecedent and consequent parts of the fuzzy rule-based systems are represented by more than one layer as will be detailed in the next subsections. Certain layers

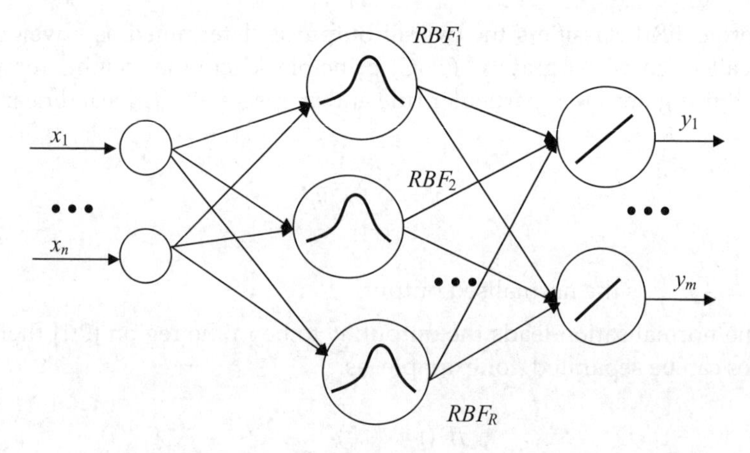

Figure 4.5 RBF-type NN that can be interpreted as a FRB

perform only multiplication or summation tasks. In general, the architecture of NFS is of feed-forward type. Perhaps, the first NFS architecture was the TS type NFS proposed by Jang (1993), which is discussed in the next subsection.

4.5.1.1 TS Type NFS

The TS type NFS can be represented as a five-layer neural network (Jang, 1993):

The functioning of the TS type NFS is as follows (see also Figure 4.6):

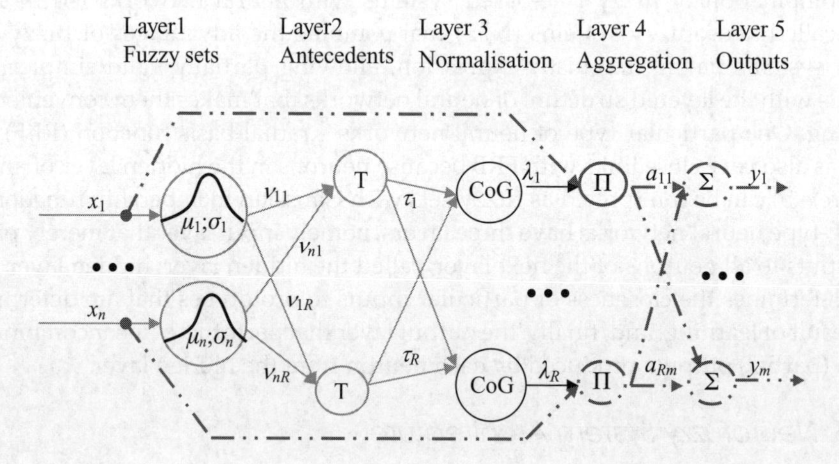

Figure 4.6 TS-type NFS architecture (the arrows that correspond to the antecedent part are represented by a solid line; the arrows that corresponds to the consequent part are shown by a dash-dotted line)

Layer 1: Every node in this layer performs the following function:

$$L_{1i} = \nu_{A_i}(x) \tag{4.27}$$

where x is the input to node i, and A_i is the linguistic label (***small, large,*** etc.) associated with this node function.

In other words, L_{1i} is the value of membership function of A_i and it specifies the degree to which given x satisfies the qualifier A_i. Usually, the preferred choice of a type of a membership function, $\nu_{A_i}(x)$ is the Gaussian one.

Layer 2: Every node in this layer is labelled by T that stands for 't-norm' and it represents the t-norm (multiplication or minimum) that is also a representation of the logical AND:

$$L_{2i} = \tau_i = \prod_{j=1}^{n} L_{1j}(x) = \prod_{j=1}^{n} \nu_j(x)$$

or

$$L_{2i} = \tau_i = \min_{j=1}^{n} \left(L_{1j}(x)\right) = \min_{j=1}^{n} \left(\nu_j(x)\right) \tag{4.28}$$

Each node output represents the firing strength of a rule.

Layer 3: Every node in this layer is a node labelled N (normalisation). The ith node calculates the ratio of the ith rule's firing strength to the sum of all rules' firing strengths:

$$\lambda_i = \frac{L_{2i}}{\sum_{j=1}^{R} L_{2j}} = \frac{\tau_i}{\sum_{j=1}^{R} \tau_{ij}}; \quad i = [1, R] \tag{4.29}$$

For convenience, outputs of this layer will be called *normalised firing strengths* (activation level).

Layer 4: Every node, i in this layer is a node with a function

$$L_{4i} = \lambda_i y_i = L_{3i} y_i = L_{3i} \left(\sum_{j=0}^{n} a_{ij} X_j\right) = \lambda_i \left(\sum_{j=0}^{n} a_{ij} X_j\right) \tag{4.30}$$

where λ_i is the output of layer 3, and a is the vector of parameters.

We will refer to these parameters as *consequent* parameters.

Layer 5: The single node in this layer is a node labelled $\sum$ that computes the overall output as the summation of all incoming signals, that is:

$$L_{5i} = \sum_{j=1}^{R} L_{4j}y = \sum_{j=1}^{R} \lambda_j y_j \tag{4.31}$$

4.5.1.2 AnYa Type NFS

The new AnYa type of FRB can also be represented by a NFS similar to the TS type one, but with one layer less (layer 2) and no parameters that needs to be learned in the antecedent part, Figure 4.7.

The functioning of the AnYa type NFS is similar to that of TS type NFS, but simpler:

Layer 1 performs the same function as in TS type NFS, however, the parameters, σ and μ (two per input dimension, that is $2n$ in total) are not necessary to determine through learning.

Layer 2 of the TS type NFS is not necessary in AnYa-type NFS at all.

Layer 2 of the AnYa-type NFS performs the same type of operation (normalisation) as **Layer 3** of the TS type-NFS, but it is now over local density to a data cloud, (4.14).

Layers 3 and **4** of the AnYa-type NFS are the same as Layers 4 and 5 of TS type NFS, respectively.

4.5.2 Evolving NFS

The evolving versions of fuzzy systems of TS type by Angelov and Buswell (2001, 2002); Angelov and Filev (2002, 2003, 2004) and of NFS (Kasabov, 2002, 2006a and b)

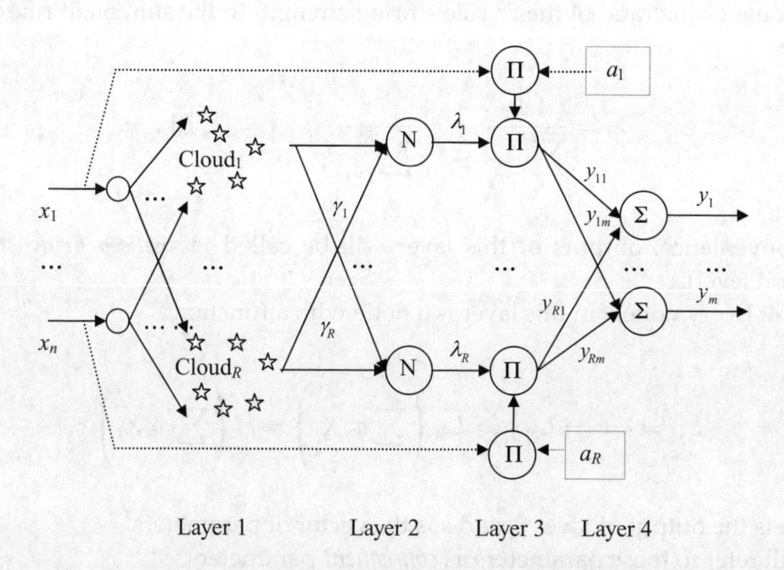

Figure 4.7 AnYa-type NFS architecture

were introduced around the turn of the centuries as techniques for online learning and autonomous adaptation of both the structure (rule-base, fuzzy sets or neurons) as well as parameters of the respective models.

Evolving fuzzy models, and, in particular, evolving TS type models, eTS (Angelov and Filev, 2004) and the more simple and advanced AnYa type (Angelov and Yager, 2012) are particularly well suited to be the framework for autonomous learning and knowledge extraction from data streams for several reasons.

The basic idea of evolving fuzzy models, EFM (Angelov, 2002) is to assume the fuzzy rule-based system to have an open (expandable or reducible) structure that adapts to the (possibly) changing data pattern similarly to the adaptation of the parameters of any type of adaptive models/systems. The difference in parameter adaptation and structure evolution is the pace – parameter adjustment is usually done at each time step (for each new data sample) while model/system structure that normally reflects the structural changes of the data pattern (appearance of a new operating regime, new state, new area of data space or disappearance of such one) usually is much more rare and, respectively, the pace of this evolution is much slower (a change of the model/system structure usually takes place once for several dozens or even hundreds or thousands of data samples).

The main driver for model/system structure evolution is the variation in the data density (pattern), while the main driver for parameter adjustment is the 'fine tuning' of an adopted structure to the particular (recent) data. Therefore, the issue of structure versus parameter adaptation is also closely related to the problem of long-term and short-term learning and to global and local (in terms of time) validity of the models. It should be noted, however, that local and global models are also considered in terms of data space (which will be discussed in more detail in Chapter 6).

4.5.3 Linguistic Interpretation of the NFS

As it was mentioned earlier, the main advantage of NFS is their linguistic interpretability (due to the link to FRB). In particular, AnYa-type NFS can also be represented by a fuzzy rule-based system as shown in Figure 4.8.

The structure, generally, can be fixed or evolving (expanding or shrinking in terms of rules/neurons and inputs/features). In many applications it is an attractive feature that allows human operators to better understand the way the system works and makes it a *'grey box'* type of system rather than a *'black box'* one as the 'pure' neural network are. It also allows both 'extraction' of human interpretable knowledge and use of such knowledge if it exists or is easy to obtain.

4.6 State Space Perspective

The AnYa-type fuzzy rule-based system and NFS can also be considered through the prism of state space representation (Kailath *et al.*, 2000). Let us have the state vector, ζ and the vector of measurable inputs x and outputs, y to the system, Figure 4.9.

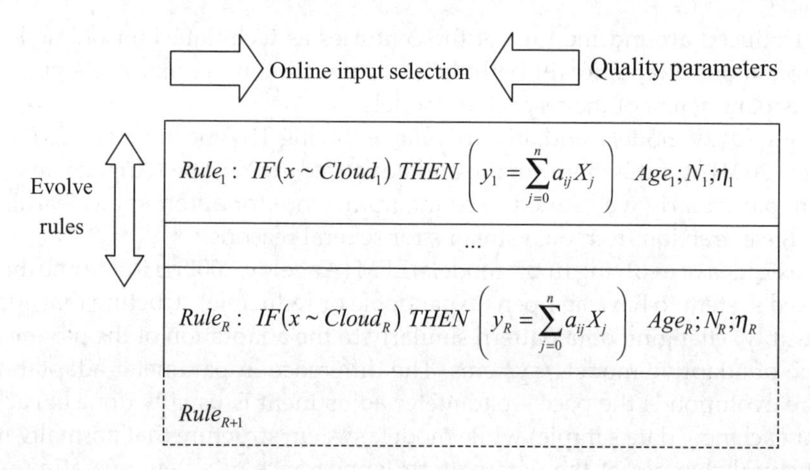

Figure 4.8 Linguistic interpretation of AnYa-type NFS – a FRB with parameter-free antecedents and linear consequents; Age_i denotes the age of the ith local submodel (fuzzy rule); N_i denotes the support of the ith local submodel (fuzzy rule); η_i denotes the utility of the ith local submodel (fuzzy rule) which will be introduced later

The state space expressions of the dynamics then are well-known (Astroem and Wittenmark, 1989) and can be represented as:

$$\zeta_{k+1} = f(\zeta_k, x_k) + w_k \tag{4.32}$$

$$y_{k+1} = h(\zeta_k, A_k) + \varpi_k \tag{4.33}$$

where w and ϖ denote the noise in the input and output channel, respectively.

In AnYa-type systems the state vector takes the form of the density, D or more specifically, the mean, μ and the scalar product, Σ through which we determine D:

$$\zeta = \begin{bmatrix} \mu \\ \Sigma \end{bmatrix} \tag{4.34}$$

Therefore, Equation (4.32) is represented by the Equations (2.31) and (2.32) that describe the update of the state vector. Equation (4.33) is represented by Equation (4.14). In summary, Equation (4.14) plus Equations (2.29–2.32) form the representation of AnYa in state space. It is nonlinear, in general, (because of the nature of the update

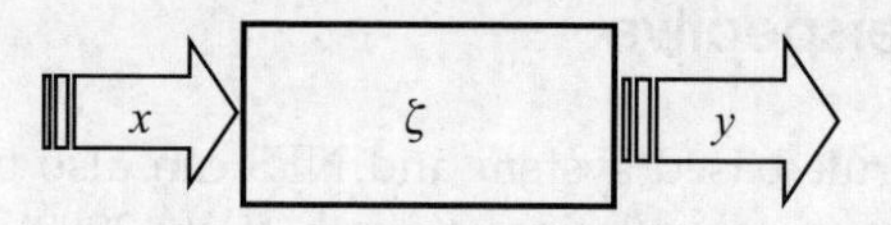

Figure 4.9 State space representation of the AnYa FRB/NFS

equations for the density and of the, generally, nonlinear nature of the model), but can be still considered and analysed as locally linear.

4.7 Conclusions

In this chapter the fundamentals of the fuzzy set theory were briefly outlined. The aim was not to provide much detail that is available in other books (Dubois, Prade and Lang, 1990; Yager and Filev, 1994), but to briefly describe the notions that will be necessary for the proposed approach. In particular, an entirely new type of FRB that offers a simpler, yet powerful description if compared with traditional Mamdani- and TS-type FRB systems has been described. FRB systems are also described in this chapter as a NFS, including the new, AnYa type and indicated what an evolving structure FRB systems (respectively, NFS) means.

In terms of classifiers, the pioneering concept of *evolving* classifiers was described that works in a similar manner to the adaptive control systems and estimators by pairs of 'classify' and 'update' actions for each new data sample (or for these new data samples for which class label is known.

Very interesting parallels were made in this chapter with the Bayesian probabilistic models that were outlined in Chapter 2 in addition to the more familiar similarities (duality) between fuzzy rule based models (of TFS and AnYa type) to RBF-type neural networks. Note, that Bayesian inference and AnYa, fuzzy rule based system of Takagi–Sugeno or AnYa type of fuzzy rule based systems or NFS as well as RBF-type neural networks are not exactly the same, but are very similar.

PART II

Methodology of Autonomous Learning Systems

5

Evolving System Structure from Streaming Data

5.1 Defining System Structure Based on *Prior* Knowledge

Traditionally, the system structure (whichever type it is, e.g. based on probabilistic models, neural networks, fuzzy rule-based systems, polynomial models, etc.) is being predefined and fixed. The choice of the structure (Bayesian, hidden Markov models, number of states, neurons, rules, order of the polynomial, type of distributions, activation functions, membership functions, number of inputs/features, etc.) is usually based on *prior* knowledge and insight from the problem domain. For this reason, such an approach is problem-specific, expert-dependent (therefore, not suitable for autonomous and online behaviour) and ignores the possible dynamic evolution of the problem at hand. The last factor becomes more and more important nowadays.

The innovative approach on which this book and the author's research in the last decade is based is focused on the development of objective, automatic, adaptive and autonomous methods for system structure identification (in a dynamic context) from the data streams. This approach was called *evolving systems* a decade ago (Angelov, 2002).

For example, initially, fuzzy systems structure identification was based on the use of *prior* expert knowledge (Zadeh, 1975; Driankov, Hellendoorn and Reinfrank, 1993). This was a logical step since the introduction of the fuzzy set theory and fuzzy linguistic variable by L. A. Zadeh was seen primarily as a tool and technique to capture and represent knowledge and linguistic (qualitative) information. To a certain extent, this approach is similar to the system structure identification based on first principles as opposed to the so-called 'black-box' type of methods (Ljung, 1987).

Both approaches rely heavily on the human involvement and lead to highly interpretable structures. Indeed, the strength of the fuzzy systems is in their ability to formalise and use human-intelligible knowledge and reasoning in the form of fuzzy rules over fuzzy sets. In this approach, expert knowledge available *a priori* is expressed verbally in a form of fuzzy rules and formalised through fuzzy sets, their membership functions, and parameters. Parameters can be further tuned by a supervised learning – an approach that will be described in the next chapter.

The adjective '*prior*' to the knowledge means that this precedes the process of system identification itself. Contrast this to the knowledge extracted from the data (*posterior* knowledge), which will be discussed later.

The structure of a system, generally, includes:

A. the number and nature of the subsystems (components);
B. their interconnection (links, overlap);
C. the number and nature of the inputs and output(s);
D. the inference mechanism.

For example, for a hidden Markov model (HMM) that means, the number of states (observable and hidden), transition probabilities, number and nature of inputs and outputs, the inference mechanism which is Bayesian.

For the example of a neural network that includes the specific types such as RBF, LVQ and so on, the number (and type) of layers, inputs and outputs, activation functions.

For the example of a fuzzy rule-based system that includes the type of the system (Mamdani, Takagi–Sugeno or AnYa), number of fuzzy rules, input and output variables, and the inference mechanism that is directly linked to the type of the fuzzy rule-based system considered. In addition, for Mamdani- and Takagi–Sugeno-type FRB this also includes the fuzzy sets (linguistic variables), linguistic connectives (aggregating operators such as t-norms and conorms).

When all of the above are predetermined by an expert (decision maker) we have a system structure identified from the *prior* knowledge.

5.2 Data Space Partitioning

An alternative to the expert *prior* knowledge approach for system structure identification is through partitioning the data space into overlapping local regions of the data space (Sugeno and Kang, 1988; Harris, 1994; Wang, 1994). The idea of this approach is straightforward – the problem of design of complex, nonlinear systems is decomposed into simpler locally valid subsystems following the millennia old Latin principle *divide et impera*.

This principle is behind such approaches as the multiple model systems (Boukhris, Mourot and Ragot, 1999), neural networks (Rumelhart and McClelland, 1986), hidden Markov models (Bishop, 2009), FRB (Sugeno and Kang, 1988), and so on. For example, the fuzzy logic is particularly suitable to formalise the flexibility and uncertainty that arises from the interconnection of the local subsystems. Fuzzy sets theory provides a convenient tool for a formal description of the transition from one local region to

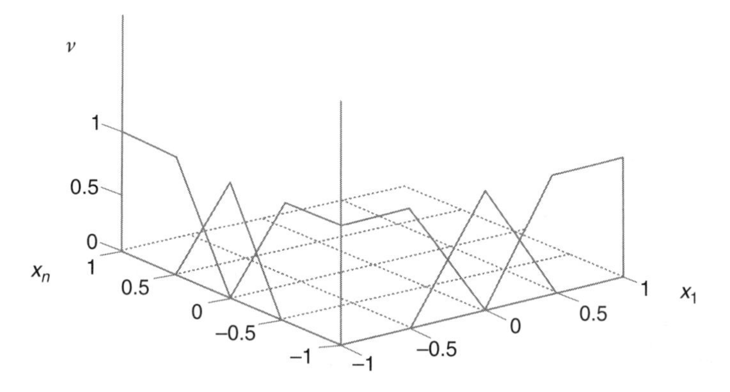

Figure 5.1 Regular orthogonal data space partitioning. The third dimension (the vertical one) represents the membership function (MF), ν if assume FRB with triangular-type MF

another one. It also provides an excellent vehicle for combination and expression of the *prior* knowledge that originates from human experts and the *posterior* knowledge that has been extracted from data patterns.

5.2.1 Regular Partitioning of the Data Space

The n-dimensional inputs data space can be evenly partitioned into a lattice of axes-orthogonal hyperboxes along the linguistic variables and the membership functions that describe them in the antecedent part of a fuzzy rule-based or a neurofuzzy system. This form of partitioning (Figure 5.1), however, requires a very large number of fuzzy rules in order to cover the whole data space. This number depends exponentially on n and on the number of linguistic terms of each variable, p:

$$R = p^n$$

For example, for a problem with two hundred inputs ($n = 200$) and five linguistic variables (e.g. *Very Low, Low, Medium, High, Very High*) for each one we get $R = 5^{200}$, which is practically intractable.

5.2.2 Data Space Partitioning through Clustering

In the more general case, the data space can be partitioned by clustering, see Section 3.2 (Chiu, 1994; Babuska, 1998; Angelov, 2004a). The shape of the clusters (Section 3.2.1) depends on the type of proximity measures used, for example, the traditional and more straightforward Euclidean distance measure leads to hyperspherical clusters (see Figure 3.4), while the more involved Mahalonobis-type distance that requires

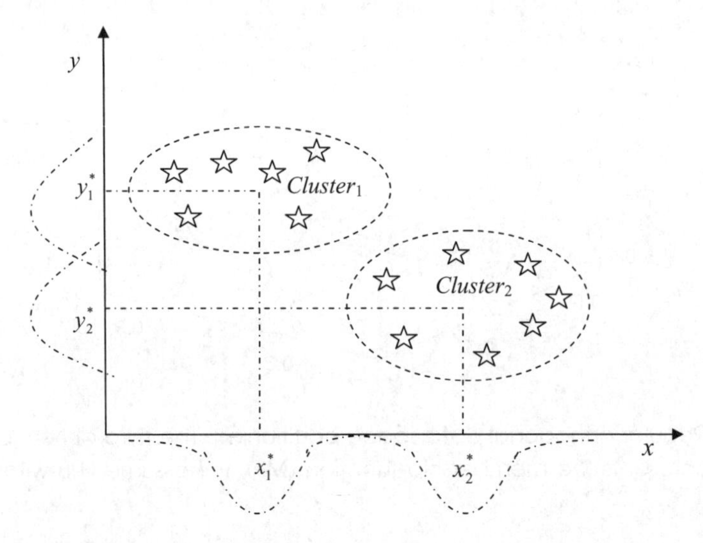

Figure 5.2 An example of hyperellipsoidal clusters resulting from the use of a Mahalonobis-distance metric. The pdfs (or respectively, membership functions) are represented by dash-dotted lines assuming Gaussian distribution

an inverse covariance matrix to be calculated leads to the more realistic hyperellipsoidal shape clusters (Figure 5.2). The reason is that the actual variance of the data is taken into account by the Mahalonobis-type distance. The resulted local regions of data subspace, generally, are described by multivariate membership (or, respectively, probability density) functions. The blending of local submodels into an overall model can be addressed either by the 'winner take all' principle, as this is done in the probabilistic context or by the 'centre of gravity' principle that tolerates *'fuzziness'* in the sense of joint and partial membership to more than one local submodel at the same time (the total membership has to be normalised so that it sums up to 1).

5.2.3 Data Space Partitioning Based on Data Clouds

In the AnYa approach the local submodels are referred to respective data subsets and not to specific subregions of the data space (e.g. clusters, regular hyperrectangles, etc.), see Figure 4.3 and compare this with Figure 5.2. In this way, no specific shape or parameters of the antecedents are defined.

The way partitioning into data *clouds* practically works is very similar to clustering (e.g. *AutoCluster* and ELM) with the main difference that there is no need to determine the centre (mean) and the radius of the data cloud. In addition, the inference of the overall model is different with the normalised firing levels (weights), λ formed in a simpler way through the local densities directly – see Equation (4.14) (Angelov and Yager, 2010).

5.2.4 Importance of Partitioning the Joint Input–Output Data Space

An important property of any partitioning of the data space for the purposes of system structure identification is the *coverage* of the whole data space. If the whole data space is not covered, situations may arise when no output can be generated, which is unacceptable and may be unsafe. This property is called *'completeness'*. One way to guarantee *completeness* is by considering the full range of all variables and apply regular partitioning assigning a local submodel to each *'cell'* of the divided data space in such a way that at least one local submodel is active (with nonzero degree). Another alternative is to use such local submodels that have guaranteed nonzero output for any input variable. For example, if we use Gaussian- or Cauchy-type density functions or membership functions, theoretically the whole data space will be covered with different from zero outputs. Alternatively, triangular-type membership functions can also guarantee completeness for a regular partitioning if there is an overlap between each neighbouring pair (see Figure 5.1).

In practice, however, data points that are outside of the so-called '3σ zone' from the focal point of the local submodel (where σ denotes the standard deviation) the value of the output produced will be negligible. Therefore, a good method for data space partitioning has to pay attention to an effective *coverage* of the whole data space to avoid computational problems (possible divisions by small numbers, which may lead to singularities).

One fundamental issue is the dimensionality of the data space that is being partitioned. In some works, the partitioning was performed per individual variable (per individual feature in classification case), for example in (Pedrycz, 1994) for so-called entropy equalisation, in (Baldwin, Martin and Pilsworth, 1995) for so-called mass assignment, in (Pedrycz, 1993) and (Berenji and Khedkar, 1993) for antecedents identification. In order to represent the complexity of interactions between all the variables, however, one needs to partition the *joint* input space of *all* input variables together. The approach to the system identification in the vector inputs (features) space (Kasabov, 1998, 2001; Angelov, 2004a) is along the lines of the VQ.

Despite being a significant step forward, vector partitioning of the input data space *only* does not represent the whole complexity of the system identification problem. A complex system is modelled by a mapping of a n-dimensional vector of input variables (in the case of classification, features) onto a (generally) m-dimensional space of outputs (in classification these might be class labels or possibility/confidence degree to a certain class). If we identify separately the input and the output data subspaces the resulting system model will not necessarily represent the correct mapping.

For example, in a simplified 2D case (single input, x and single output, y) that is represented in Figures 5.3 and 5.4, if we analyse the input data space only a single model will suffice. If, however, the joint input–output data space is considered, it is obvious that two separate submodels will be necessary. For example, if the data in one of the local submodels suggests output *'Turn Left'* for a mobile robot while the other local submodel suggests *'Turn Right'*, having only one model will practically

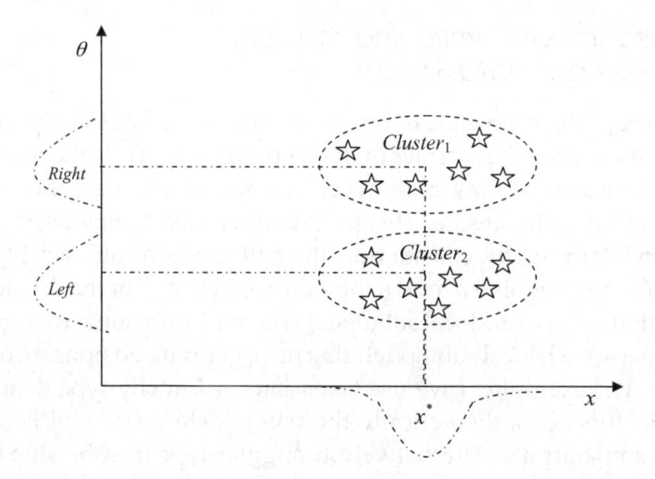

Figure 5.3 In terms of the input, x only one local model appears to be needed, but two obviously different outputs are produced (respectively, suggesting to turn left or right)

result in averaging all the data and, as a result, the output may be *'Go Straight'*, which can have damaging consequences. In this case, the two local models aggregated using 'winner takes all' (or 'few winners take all' in the case of more than two alternatives) type of inference instead of the typically used 'centre of gravity' one would avoid the problems described.

This simple example demonstrates that the joint input–output $(n+m)$-dimensional data space has to be considered rather than the n-dimensional input and m-dimensional output data subspaces separately. When normalisation to the range [0;1] is applied the modelling is a mapping $x \in [0;1]^n \rightarrow y \in [0;1]^m$.

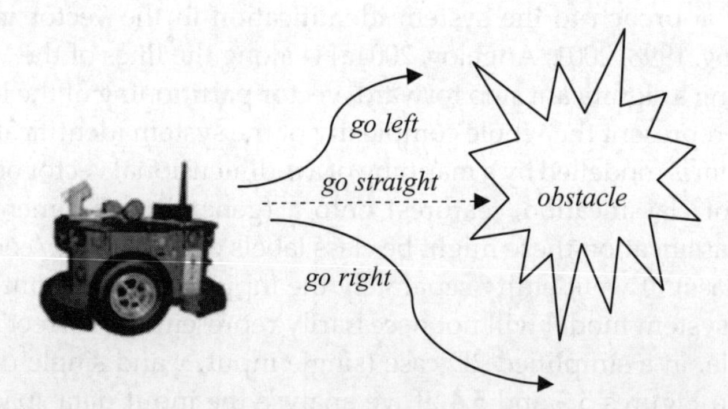

Figure 5.4 A simplified 2D example (a 'robot' and an obstacle) demonstrating the problem that may occur if we consider input and output data subspaces separately. The dashed line in the middle represents 'go straight', which is an average resulting from Figure 5.3

5.2.5 Principles of Data Space Partitioning for Autonomous Machine Learning

The proposed methodology for autonomous machine learning is based on the following principles for decomposition of the data space into (possibly overlapping) local regions:

A. new data samples that have high descriptive power (e.g. estimated by its density) are eligible to be added as focal points of new local regions (submodels);

$$\begin{aligned}&\text{IF } \left(D(x_k) > \max_{i=1}^{R} D(x_i^*)\right)\\&\text{THEN } \left(x_k \rightarrow x_{R+1}^*\right)\end{aligned}$$

B. new areas of data space that were not covered before should be covered if they contain a certain number of new data samples;

$$\begin{aligned}&\text{IF } \left(\delta_{\min} = \min_{i=1}^{R} \left\|x_k - x_i^*\right\|^2 > \frac{r}{2}\right)\\&\text{THEN } \left(x_k \rightarrow x_{R+1}^*\right)\end{aligned}$$

C. Focal points of local regions (submodels) should not:

i. be very close to each other, to avoid significant overlap;

$$\begin{aligned}&\text{IF } \left(\delta_{\min} = \min_{i=1}^{R} \left\|x_k - x_i^*\right\|^2 \leq \frac{r}{2}\right)\\&\text{THEN } \left(x_k \rightarrow x_{j^*}^*;\ j^* = \arg\min_{i=1}^{R} \left\|x_k - x_i^*\right\|^2\right)\end{aligned}$$

ii. describe *old* data only, see Equation (5.12);
iii. have *low utility*, see Equation (5.14).

5.2.6 Dynamic Data Space Partitioning – Evolving System Structure Autonomously – example of Fish Classification

Methods for data space partitioning described above (using *prior* knowledge, using regular partitioning, using clustering or data clouds) can be done offline, in a batch mode, but all except the first one (using *prior* knowledge) can be done in the online mode, recursively also. This is in the centre of the concept of evolving systems – to extract the system structure automatically online from the data stream. In this way, the overall system is a (fuzzy) blend of a much simplified partially, locally simpler (possibly, linear) submodels that are **not prefixed or predetermined**, but rather adapt/evolve based on the changing data density pattern following the main principles listed above.

Let us consider a simple standard classification problem. For example, let us try to classify fish into two categories

a. 'Salmon'; and
b. 'Sea Bass'.

The features based on which an automatic system can categorise them can be, for example, the size (length) and weight of the fish.

For this simple example one can consider all types of data space partitioning assuming a Bayesian probabilistic model as well as a fuzzy rule-based type of model. Let all the data samples that are available from both types of fish be represented in Figure 5.5.

5.2.6.1 Data Space Partitioning Based on Prior Knowledge

For this simple problem one can assume the following data space partitioning (Figure 5.6):

- Area 1 is typical for Salmon (length >1.5 m; weight >1.5 kg);
- Area 2 is typical for Sea Bass (length <1.5 m; weight <1.5 kg).

However, it is well known and also obvious from Figure 5.5 that some species of Salmon can be smaller or lighter and on the contrary, some Bass can be larger or

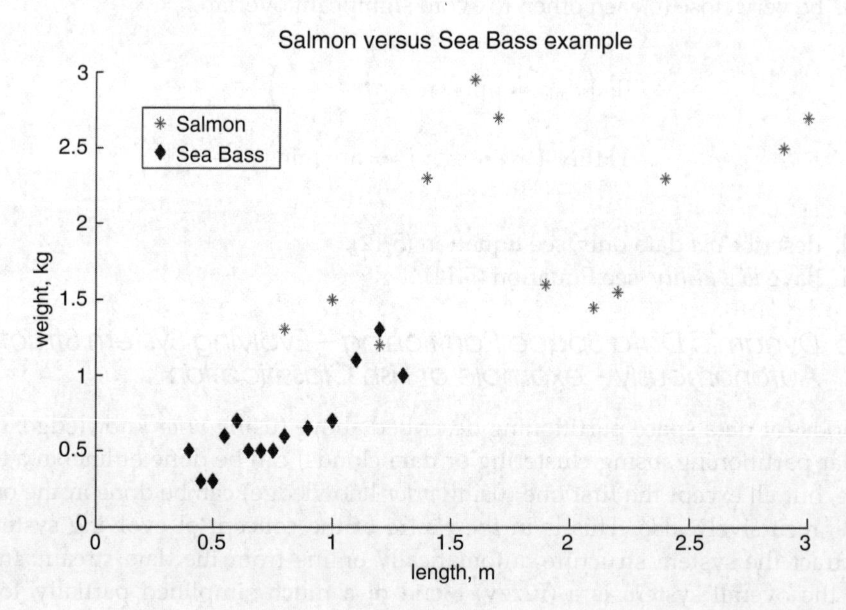

Figure 5.5 Data from two types of fish regarding two features – length and weight (stars denote *Salmon*; diamonds denote *Sea Bass*)

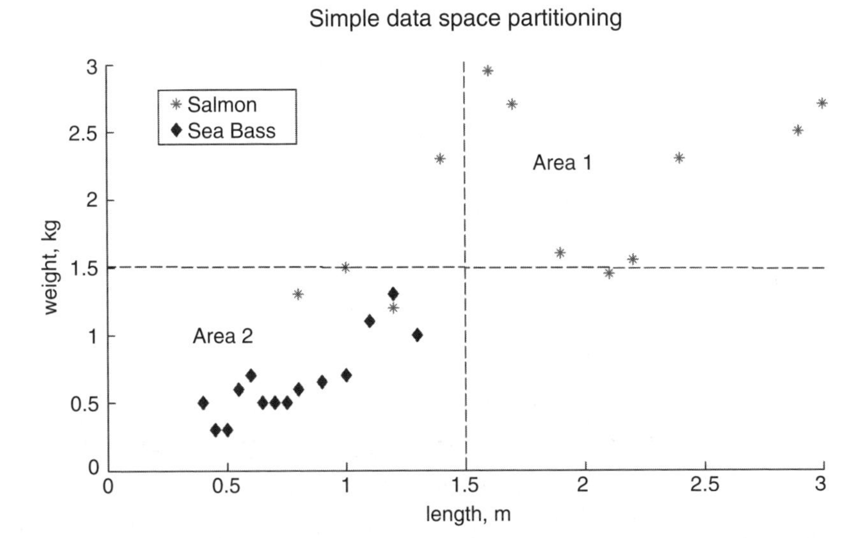

Figure 5.6 Data space partitioning based on prior knowledge for the example of two types of fish represented in Figure 5.5

heavier. So, partitioning based on *prior* knowledge represents some form of accumulated and averaged *prior* knowledge and principles rather than the specific situation and data at hand.

One can develop, instead, local submodels (e.g. Bayesian or FRB) for the two Areas. For example, the Bayesian model can be developed based on the pdf of the data in this Area. FRB model can have the form of Mamdani-type fuzzy rules:

$$\begin{aligned} R_1&: IF\ (Data\ sample\ is\ in\ Area1) \\ &\quad THEN\ (Salmon) \\ R_2&: IF\ (Data\ sample\ is\ in\ Area2) \\ &\quad THEN\ (Sea\ Bream) \end{aligned} \tag{5.1}$$

where (*Area is* 1) is the antecedent represented by the fuzzy sets ($Length > 1.5$ m) AND ($Weight > 1.5$ kg), which themselves are represented by membership functions.

For example, membership functions of Gaussian type for a fuzzy description of the variable length of the fish are depicted in Figure 5.7.

Another example of membership functions of the same variable of triangular type is depicted in Figure 5.8.

It should be stressed that the membership functions may look very similar to the pdfs, but the main difference is that they represent a degree of membership to the fuzzy set, not the probability and the maximum of the membership functions is 1 because the membership function do not satisfy the condition (2.23). The main

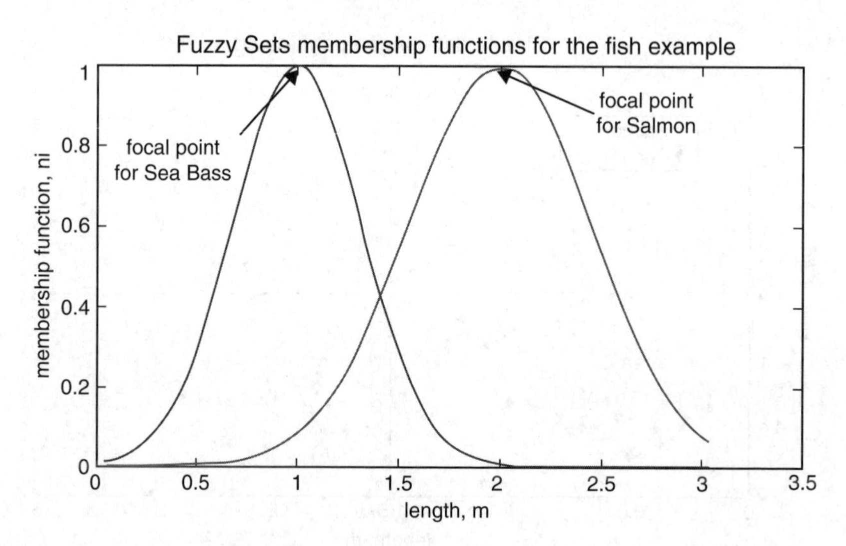

Figure 5.7 An example of Gaussian-type membership functions for a fuzzy description of the variable length of the fish

problems (for both FRB and Bayesian type of representation) is, however, that the forms convenient for mathematical treatment such as normal Gaussian distributions are not necessarily adequately describing the real data, as can be appreciated from Figures 5.5 and 5.6. Alternatives include, in the probabilistic domain, so-called particle filters (Arulampalam, Maskell and Gordon, 2002; Doucet, Godsill and Andrieu,

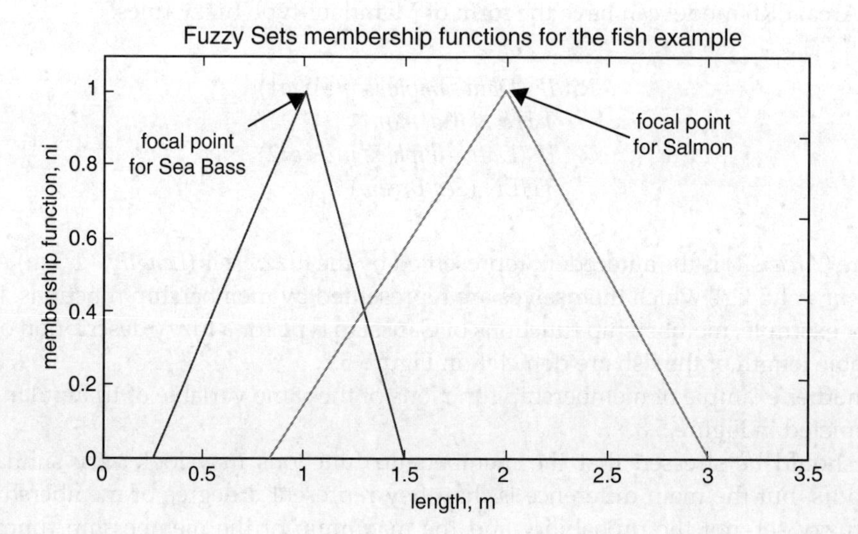

Figure 5.8 Triangular-type membership functions for a fuzzy description of the variable length of the fish

2000) and in the area of FRB systems the AnYa type of representations and using of data clouds.

5.2.6.2 Regular Data Space Partitioning

Regular data partitioning (Figure 5.6) is similar to the partitioning based on *prior* knowledge in the sense that it also disregards the actual data. For example, the whole range of lengths and weights for both types of fish can be divided into two equal parts and, thus, have four local regions of data space where respective local submodels (e.g. Bayesian or fuzzy) can be developed. In fact, the data are primarily in two of these four regions – see Equation (5.1) and Figure 5.6.

5.2.6.3 Data Space Partitioning through Clustering

The same data can also be separated into (possibly overlapping) clusters using any suitable clustering method, see Figure 5.9 below.

5.2.6.4 Data Space Partitioning through Clouds

Finally, if we use data *clouds* one only needs to identify the focal points (Figure 5.10). One can, then, associate the incoming data points based on their local density

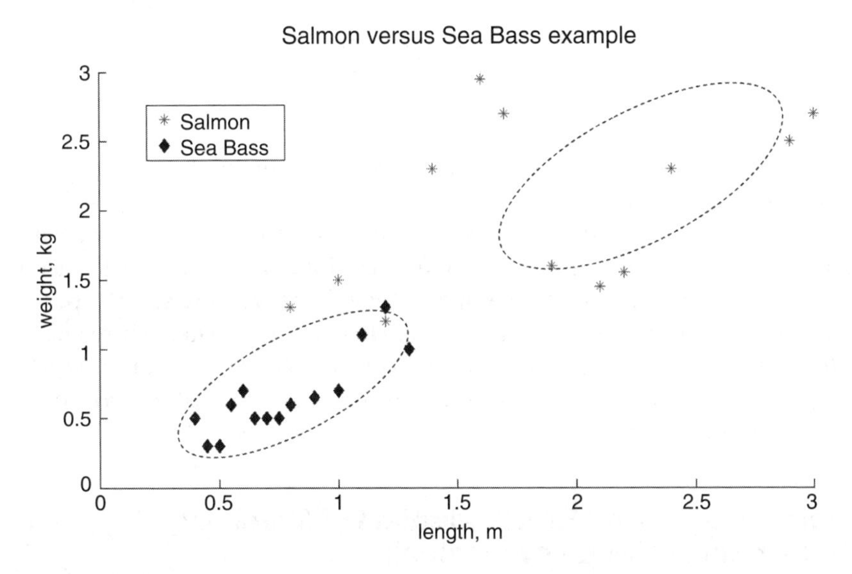

Figure 5.9 Data space partitioning through clustering for the example of two types of fish represented in Figure 5.5. The problems with using clusters (even if they are ellipsoidal – using Mahalonobis distance metrics) are obvious

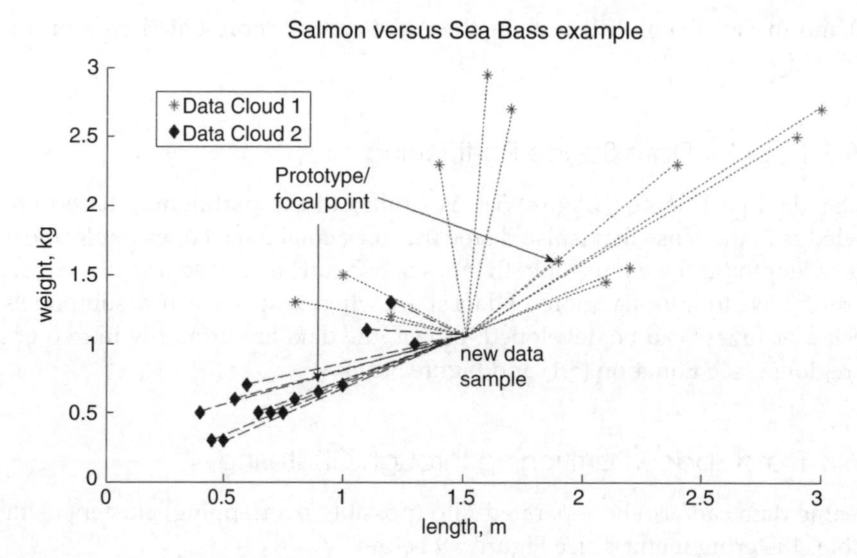

Figure 5.10 Data space partitioning through clouds for the example of two types of fish represented in Figure 5.5 (the relative density to a cloud is determined through the distances from the new data sample to all points from each cloud

(a data item can be assigned to the data cloud, which has maximum local density, Equation (5.2)).

$$z_k \in C_{j^*} \big| \; j^* = \arg\max_{i=1}^{R} (d_i) \tag{5.2}$$

where C_{j^*} denotes the j^{*th} data cloud.

Each focal point can also be a prototype for a local submodel.

In this way, the main elements of the structure of a complex system – the number and position of the local simpler (possibly, linear) submodels can be determined by one of the above ways with clustering and data clouds being the most promising ones. Data clouds have an additional advantage – they do not require specific parameters, while clustering requires two parameters per cluster – the cluster centre and radius.

In the remainder of this chapter we will consider in more detail how to evolve different aspects of the structure, such as normalisation and standardisation in an evolving environment, number of inputs, *age*, *utility* and *radii* of the local submodels.

5.3 Normalisation and Standardisation of Streaming Data in an Evolving Environments

A very important problem related to any online algorithm is the normalisation or standardisation of the streaming data. Applying one of the two techniques is necessary in order to transform the data so that they are made comparable. Both, normalisation and standardisation are well established techniques for the offline case when all the data is available in a batch mode, see Section 3.1.1. In online environment, however, the range of the data is not necessarily known in advance. Moreover, in an evolving system structure scenario new local submodels may be formed and have to be initialised and updated properly. Therefore, both normalisation and standardisation cannot be applied straightforwardly as in the offline case and alternative algorithms are required.

5.3.1 Standardisation in an Evolving Environment

In an online mode the data stream may have time-variable statistical characteristics (being, generally, a nonstationary one) and, therefore, both mean, μ and standard deviation, σ, generally, change (evolve). Although, the update of both the mean and standard deviation can be done easily, see Equation (2.31):

$$\mu_k = \frac{k-1}{k}\mu_{k-1} + \frac{1}{k}z_k \qquad \mu_1 = z_1 \tag{5.3}$$

where

$z = [x^T; y^T]^T$; $z \in R^{n+m}$ denotes the joint input–output data sample
$x = [x_1, x_2, \ldots, x_n]^T$ is the n-dimensional input vector
$y = [y_1, y_2, \ldots, y_m]$ – the m-dimensional output vector.

The standard deviation can be updated as follows (Duda, Hart and Stork, 2000), see also Equation (2.32):

$$\sigma_k^2 = \frac{k-1}{k}\sigma_{k-1}^2 + \frac{1}{k}\|z_k - \mu_k\|^2$$
$$\sigma_1^2 = 1 \tag{5.4}$$

The main difficulty in standardisation in an evolving environment is, actually, not the operation itself (Equations (5.3) and (5.4)), but the fact that this update has to be done for **every** new data sample and it does not apply retrospectively. On the other hand, the normalisation requires update **only** when new data exceeds one of the boundaries, which may practically be rare if we know these boundaries well in advance. In addition, a single data sample cannot exceed both boundaries (lower and upper) at the same time and, thus, only one of the boundaries needs to be updated sometimes. This argument is in favour of the normalisation for an evolving environment because it leads to fewer computations and is exact because it treats all data points equally, despite the fact that the range of the data may change.

Standardisation in an evolving environment, on the contrary, is asymptotical, because both mean and standard deviation are changing with each new data sample and, thus, each time the data is standardised under varying conditions. The changes to the algorithm parameters, however, which are required by the evolving normalisation, are abrupt; because any new data point that is outside of the range that applies up to this point extends it. The changes due to evolving standardisation, on the contrary, are more robust because the mean and standard deviation accumulate/filter the effects of all the data and do not change suddenly and drastically. Therefore, without limiting the choice, we recommend evolving standardisation that confines over 99% of the standardised data into the range [−3;3] (Duda, Hart and Stork, 2000).

5.3.2 Normalisation in an Evolving Environment

Zhou and Angelov (2007) proposed normalisation in an evolving environment based on the update of the minimums and maximums per input variable (see Equation (3.3)) which takes place only when the new data sample leads to one of the extreme values (*lower* or *upper* boundary of $z_j; j = 1, 2, \ldots, n$) to be exceeded.

The procedure (Zhou and Angelov, 2007) starts with the third data sample, z_k ($k = 3, 4, \ldots$), because it makes no sense for two samples or less. First, it is checked if z_k is within the current boundaries:

$$\underline{z}_k \leq z_k \leq \overline{z}_k \tag{5.5}$$

where

$\overline{z}_k = \max_{i=1}^{k-1}(z_i)$ denotes the current *upper* boundary of z at the moment of time k;

$\overline{z}_k = \min_{i=1}^{k-1}(z_i)$ denotes the current *lower* boundary of z at the moment of time k.

If Equation (5.5) is satisfied then z_k is normalised using the current lower and upper boundaries, $\underline{z}_k$ and $\overline{z}_k$ by applying Equation (3.3). If Equation (5.5) is not satisfied then, first, the boundaries, $\underline{z}_k$ and $\overline{z}_k$ are updated using Equations (5.6) or (5.7) corresponding to which part of Equation (5.5) is violated – they cannot be violated both at the same time. After that, we normalise z_k as follows:

$$\underline{z}_k = \min(\underline{z}_{k-1}, z_k) \tag{5.6}$$

$$\overline{z}_k = \max(\overline{z}_{k-1}, z_k) \tag{5.7}$$

Let us introduce the ratio $\rho_k = \frac{\overline{z}_{k-1} - \underline{z}_{k-1}}{\overline{z}_k - \underline{z}_k}$. Note, that quite often, when there is no need for change of the range, in practice, $\rho = 1$. Each time the normalisation boundaries change, all the parameters of the algorithm need to be updated with the ratio ρ as an additional parameter (Zhou and Angelov, 2007).

5.4 Autonomous Monitoring of the Structure Quality

5.4.1 Autonomous Input Variables Selection

Traditionally, the number of input variables, n is considered to be **predefined and fixed.** This is very often the case, because the factors that affect the output are, usually, known from the *prior* knowledge. In many problems, however, some input variables are highly correlated. To avoid this problem the traditional offline approach, which is a part of preprocessing, includes techniques such as orthogonalisation through PCA (Li, Yue and Valle-Cervanteset, 2000), selection of input variables (features) through techniques such as genetic programming (GP) (Kordon and Smits, 2001), partial least squares (PLS) (Fortuna *et al.*, 2007), sensitivity analysis (Hastie, Tibshirani and Friedman, 2001), and so on. PCA and PLS have two major disadvantages:

i. The model interpretation is difficult because they are, in fact, linear combinations of the original input variables; and
ii. They are limited to the linear systems. GP is very slow and computationally expensive.

It is also offline and assumes stationarity of the data; its results are valid only if the data distribution does not change for the validation data in comparison with the distribution of the training data.

In an online scenario when the system structure evolves the selection of most suitable input variables becomes even more challenging. It is order dependent and not retrospective (a certain variable may be important based on certain subset of data, but not important based on the next subset, for example). Therefore, automatic input variable selection 'on the fly', even if possible, is not necessarily optimal or unique in global sense. Such a technique allows the system initially to encompass all available inputs and automatically trim gradually the structure in terms of selecting the most important inputs only.

One approach for input selection (automatic gradual reduction of the number of used input variables) is based on the contribution each input has to the overall model (Angelov, 2006). For this purpose, the local submodels (usually, linear) are monitored online and the values of their parameters are analysed (Angelov, 2010). If the values of the parameters are negligibly small across the local submodels for certain input (feature), $j = [1, n]$ and certain output (in a MIMO case), $l = [1, m]$ for all the data samples seen so far, the algorithm removes this particular input/feature, j^* because it is not contributing significantly towards the overall output.

This can be expressed mathematically by the accumulated sum of the local submodels' parameters for the specific, j^{th} input/feature in respect to (normalised by) all n inputs/features (Angelov, 2006):

$$\omega_{ijlk} = \frac{\pi_{ijlk}}{\sum_{r=1}^{n} \pi_{irlk}}, \quad i = [1, R]; \; j = [1, n], \; l = [1, m] \tag{5.8}$$

where $\pi_{ijlk} = \sum_{r=1}^{k} |a_{ijlr}|$ denotes the accumulated sum of parameters' values.

The values of a_{ijlk} indicate the contribution of a particular input/feature that can be monitored. The values of ω_{ijlk} indicates the ratio of the contribution of a particular (j^{th}) input/feature compared with the contributions of all features, $p_{ilk} = \sum_{r=1}^{n} \pi_{irlk}$ or with the contribution of 'the most influential' feature, $\overline{\pi} = \max_{r=1}^{n} \pi_{irlk}$, $l = [1, m]$, $i = [1, R]$, $t = 1, 2, \ldots$ In Equation (5.8) the former is assumed.

An extensive empirical study was made by Angelov and Zhou (2008) which indicates that it is more appropriate to compare the ratio out of *all* features for problems with less than 10 inputs/features. At the same time, for problems with multiple (over 10 inputs/features) the same study suggests to use the ratio out of the most influential input/feature, $\overline{\pi}$. When the number of features is large the sum of contributions becomes a large number on its own and masks the effect of a particular feature. Therefore, the comparison in this case is with the feature that contributes most, $\overline{\pi}$.

The condition to automatically select the most relevant inputs/features can be summarised as follows (Angelov, 2006):

$$\begin{array}{l}\text{Condition AIS1} \quad IF\left(\exists j^* | \, \omega_{ij^*lk} < \varepsilon p_{ilk}\right) \; AND \, (n \leq 10) \; THEN \, (remove \; j^*) \\ \text{where } i = [1, R], \; l = [1, m], \; k = 2, 3, \ldots\end{array} \tag{5.9}$$

$$\begin{array}{l}\text{Condition AIS2} \quad IF\left(\exists j^* | \, \omega_{ij^*lk} < \varepsilon \overline{\pi}_{ilk}\right) \; AND \, (n > 10) \; THEN \; (remove \; j^*) \\ \text{where } i = [1, R], \; l = [1, m], \; k = 2, 3, \ldots\end{array} \tag{5.10}$$

where

AIS denotes automatic inputs selection;
ε denotes a coefficient – (suggested values are 3 to 10%).

It should be stressed that the removal of an input/feature leads to a change in the dimension (shrinking) of the overall system. Therefore, the inputs vector, x, respectively, the focal points of local submodels, x^*, covariance matrix, $\sum$, standard deviation, σ and all variables of the algorithm related to them will have a new dimension $n(k) < n(k-1)$ if the removal of the input is made at the moment of time k.

When new inputs/features are being added a new column and a line is being added in the covariance matrix and initialised in the same way as if a new rule is being added, see Section 6.3.

Let us have the matrix C_k and the ith feature is to be removed. Then the following transformation of the matrix takes place for the removal of a feature:

$$C_k = \left[\begin{array}{cc|cc} c_{11} & \ldots & c_{1i} & \ldots \; c_{1n} \\ \ldots & \ldots & \ldots & \ldots \; \ldots \\ \hline c_{i1} & \ldots & c_{ii} & \ldots \; c_{in} \\ \ldots & \ldots & \ldots & \ldots \; \ldots \\ c_{n1} & \ldots & c_{ni} & \ldots \; c_{nn} \end{array}\right]$$

$$C_{k+1} = \begin{bmatrix} c_{11} & \cdots & c_{1(i-1)} & c_{1(i+1)} & \cdots & c_{1n} \\ \cdots & \cdots & \cdots & \cdots & \cdots & \cdots \\ c_{(i-1)1} & \cdots & c_{(i-1)(i-1)} & c_{(i-1)(i+1)} & \cdots & c_{(i-1)n} \\ c_{(i+1)1} & \cdots & c_{(i+1)(i-1)} & c_{(i+1)(i+1)} & \cdots & c_{(i+1)n} \\ \cdots & \cdots & \cdots & \cdots & \cdots & \cdots \\ c_{n1} & \cdots & c_{n(i-1)} & c_{n(i+1)} & \cdots & c_{nn} \end{bmatrix}$$

and similarly, for the addition of a new feature:

$$C_k = \begin{bmatrix} c_{11} & \cdots & c_{1i} & \cdots & c_{1n} \\ \cdots & \cdots & \cdots & \cdots & \cdots \\ c_{i1} & \cdots & c_{ii} & \cdots & c_{in} \\ \cdots & \cdots & \cdots & \cdots & \cdots \\ c_{n1} & \cdots & c_{ni} & \cdots & c_{nn} \end{bmatrix}$$

$$C_{k+1} = \begin{bmatrix} c_{11} & \cdots & c_{1i} & \cdots & c_{1n} & 0 \\ \cdots & \cdots & \cdots & \cdots & \cdots & 0 \\ c_{i1} & \cdots & c_{ii} & \cdots & c_{in} & 0 \\ \cdots & \cdots & \cdots & \cdots & \cdots & 0 \\ c_{n1} & \cdots & c_{ni} & \cdots & c_{nn} & 0 \\ 0 & 0 & 0 & 0 & 0 & \Omega \end{bmatrix}$$

5.4.2 Autonomous Monitoring of the Age of the Local Submodels

The *age* of the (evolving) local submodel can be defined (Angelov, 2006) as the accumulated time of appearance of the data samples that form the cluster or cloud based on which the local submodel was formed:

$$Age_{ik} = k - \frac{\sum_{l=1}^{N_{ik}} I_l}{N_{ik}}; \; i = [1, R] \qquad (5.11)$$

where

k denotes the current time instant;

N_{ik} denotes the number of data samples (support) that are associated with the cluster or cloud;

I_l denotes the time index of the moment when the respective data sample was read.

The values of *Age* vary from *0* to k and the derivative of *Age* in respect to time is always less than or equal to *1* (Lughofer and Angelov, 2011). Clusters (clouds) are said to be *'old'* (their *Age* is high) when they have not been updated recently. *'Young'* cluster (clouds) have predominantly new samples or recent ones. The (first and second) derivatives of the *Age* are very informative and useful for detection of data *'shift'* and *'drift'* (Lughofer and Angelov, 2011). The *Age* indicates *how old* the information that supports certain local submodel is. One can monitor the *Age* of each local submodel online and compare this with the *mean Age* that is determined as $\overline{Age}_k = \frac{1}{R}\sum_{i=1}^{R} Age_{ik}$, which can also be updated online. One can use the *Age* to remove older local submodels or to detect the concept *drift* that corresponds to the inflexed point of the *Age* curve (the point when the derivative of Age in terms of time index, $\frac{d(Age)}{dk}$ changes its sign (Angelov and Kordon, 2010).

When a new data sample creates a new local submodel, its *Age* is initiated (the new cluster is 'born'). Each time a new data sample is associated with an existing local submodel, the *Age* of that cluster gets smaller/younger (see Equation (5.11)). If no sample is associated with this local submodel it gets older.

The *Age* of local submodels is especially important for data streams. It gives accumulated information about the timing ***when*** a certain sample was assigned to a cluster or cloud. It is well known that incremental approaches are order dependent. With the *Age* one can make use of this specific feature of the data streams.

The following simple rule for update of global system model can be formulated for **removing** *older* local submodels (whose *Age* is above the mean *Age* by more than 'one sigma'):

$$\begin{aligned} &IF\ \left(Age_i > \overline{Age_i} + \sigma_{Age_i}\right) \\ &THEN\ (\lambda_i \leftarrow 0; R \leftarrow (R-1)) \end{aligned} \tag{5.12}$$

where $\overline{Age}_i$ denotes the mean *age*; σ_{Age_i} denotes the standard deviation of the *Age* of the ith rule.

5.4.3 Autonomous Monitoring of the Utility of the Local Submodels

The *utility* of the local submodel was introduced for fuzzy rule-based type systems (Angelov, 2006) as the degree of support for this local submodel by the future data samples. *Utility* can be seen as a measure of the support for the local submodel (it takes real numbers in the range [0;1] while the *Age*, for example, takes crisp integer values). *Utility* is a more representative measure, because it is based on the weight (firing level, confidence) rather than on the distance to the focal point only as in the case of *Age* or frequency as in pdf. Utility is formulated as (Angelov, 2006):

$$\eta_{ik} = \frac{\sum_{l=1}^{k} \lambda_l}{k - t_i};\ i = [1, R] \tag{5.13}$$

where

t_i denotes the time instant when the ith local submodel has been created;

λ denotes the firing level (in the case of a fuzzy rule), confidence level (in the case of a probabilistic model), weight or activation function.

Utility, η_i, accumulates the weight of the local submodel's contributions to the overall output during the life of that local submodel (from the current time instant back to the moment when this local submodel was generated). It is a measure of importance of the respective local submodel. The relative nature of λ (see Equation (3.49) or, respectively, Equation (4.14) of the centre of gravity inference mechanism) holds the comparison between a given local submodel and all other local submodels.

Similarly to the case with the *Age*, *utility* provides a tool to benefit from the order dependency of the models that are derived from data streams. These tools can be used to address the nonstationary nature of the data streams by evolving/adapting the structure of multimodel systems in terms of number of local submodels, but also in terms of their relevance once created measured by the *Age* and *utility*. A similar rule to Equation (5.12) for removal of local submodels with low *utility* can be formulated:

$$\begin{array}{l} IF\ (\eta_i < \varepsilon) \\ THEN\ (\lambda_i \leftarrow 0;\ R \leftarrow (R-1)) \end{array} \tag{5.14}$$

where ε denotes a coefficient – (suggested values are 3 to 10%).

5.4.4 Update of the Cluster Radii

The value of the cluster radii (zone of influence of the clusters), r can also be updated online in a data-driven fashion by learning the data distribution and variance (Angelov, 2006):

$$r_{ijk}^2 = \alpha r_{ijk-1}^2 + (1-\alpha)\frac{1}{N_{ik}}\left(z_{jk} - z_j^*\right)^2 \tag{5.15}$$

where

α denotes the learning step (recommended value 0.5);

N_{ik} denotes the number of data samples that are associated with the ith cluster based on the closeness; the initial value of the spread is usually $r_{j1} = 0.5$.

The cluster radius, r is an important parameter that affects the results. For example, in fuzzy rule-based systems it is a part of the membership function and, thus, of the activation level of the fuzzy sets; in Bayesian pdf it determines the spread of the Gaussian and, in general, it determines the zone of influence of the cluster. Traditionally, it is assumed to be predefined, e.g. in Mountain (Yager and Filev, 1994) and subtractive clustering (Chiu, 1994), even in early versions of *AutoCluster* (Angelov, 2004a).

It has to be noted that the absence of problem-specific parameters is an obvious advantage of any algorithm (this is especially true for online algorithms). While the parameter α is suitable for all problems, the value of the radius, r is, to a great extent, problem specific (although suggested values in range [0.3;0.5] for normalised data is

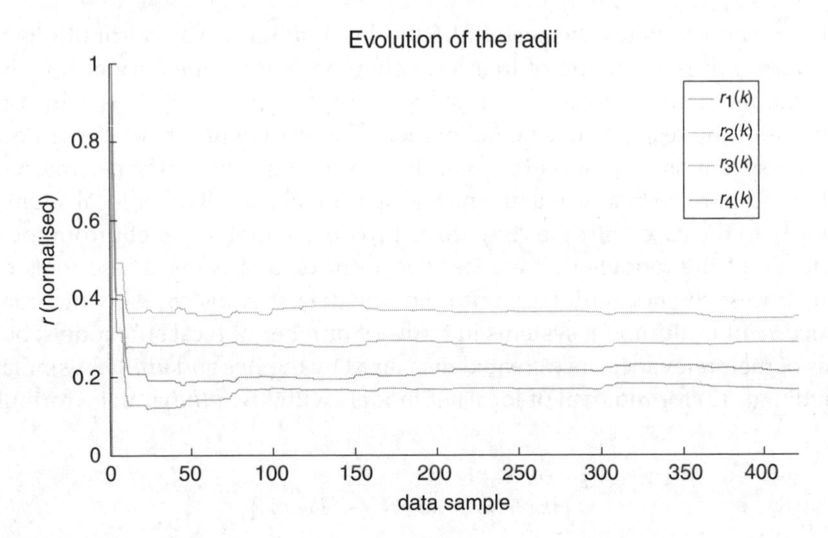

Figure 5.11 Evolution of the radii of the clusters – an example based on Equation (5.15)

also relatively problem independent, it can be demonstrated that it does affect the results). Figure 5.11 illustrates the evolution of the radius.

From Figure 5.11 one can see that the radii of all the clusters converge to a stable value pretty quickly.

5.5 Short- and Long-Term Focal Points and Submodels

The system is adapted (it evolves) as a reaction to significant structural changes that are marked by innovations. These innovations are represented by data samples that are not described well by the existing structure (including all subsystems). In this respect, innovations are identical to anomalies (see Section 2.6).

However, if a new submodel is established for each anomaly, the model quickly becomes intractable, complex and biased, not reflecting correctly the total data distribution (on average). This is the main disadvantage of some greedy clustering algorithms and the evolving systems based on them (Leng, McGuinty and Prasad, 2005) that later require some sort of '*pruning*'.

Instead, a more prudent and robust approach is to require the density of the new focal points to be high (as *AutoCluster* does), see principle A from Section 5.2.5. When new focal point is established in accordance with the principle B (same section), that is, to cover the data space, an additional condition must cover the minimum number of points in the vicinity (e.g. 3) or local density. In this book a new approach to evolve system structure (by considering two types of focal points and, respectively, local submodels) is proposed. This applies not only to fuzzy rule-based systems, but also to neural networks, probabilistic multimodel systems. To the best of the author's knowledge, none of the evolving systems (starting from the very first, such as eTS and DENFIS which are dated back to the beginning of the century) as well as non-evolving adaptive and offline systems, including probabilistic multimodel systems, neural network and fuzzy rule bases use this method.

The idea is to establish any innovation (which initially is nothing more than an outlier) as a candidate focal point of a local submodel and to declare and use it as a regular one only after it is supported by more points in its vicinity. In this way, it is possible to differentiate between short-term memory that keeps all candidate focal points (outliers) and long-term memory that keeps only focal points that are actually used in the model. Of course, focal points that are created based on the principle A (see Section 5.2.5) are always used directly unless they contradict principle C, item i).

In addition, principle C (items ii) and iii)) takes care at a later stage of removing the submodels that are based on outliers (short-term memory), because they will be either not used further (low *utility*) or *old* (no new data samples will be associated with them).

5.6 Simplification and Interpretability Issues

Selecting the focal points and designing local submodels based on them follows the general principles described in Section 5.2.5. Although, the principle C i) takes care of data samples that are close to an existing focal point not to be considered as a new one it applies in a vector sense (multidimensional distance), see Figure 5.12.

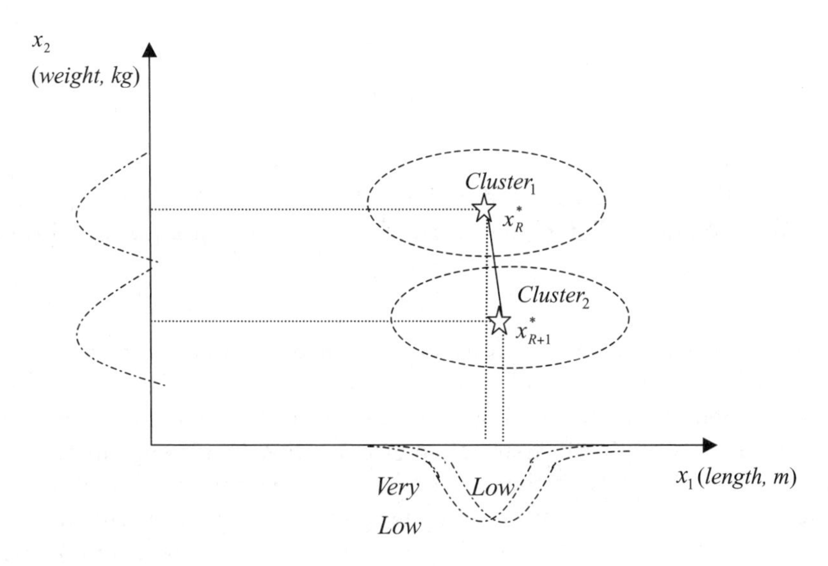

Figure 5.12 The two focal points are not close to each other overall (in a vector sense), but are close to each other in terms of one of the dimensions

Such data samples (like the one depicted in Figure 5.12) for which the distance in one (or few) particular dimensions are very close to previously existing focal points can lead to interpretability problems (ambiguity). This is a particularly important and acute problem for the fuzzy rule-based systems (which pride themselves precisely for the interpretability, transparency and human intelligibility), but is also a problem for probabilistic, neural network systems and so on.

For example, the new focal point may have been established following principles A and C i) – see Section 5.2.5; however, in terms of one particular dimension (e.g. 'length of the fish') it may have almost the same position and, therefore, parameters as the existing focal point x_R^*.

In terms of FRB this means that the linguistic variables that will be related to the fuzzy sets with respect to the variable *length* will have approximately the same meaning although different linguistic label (*Low* and *Very Low*). In terms of a probabilistic model this will mean that the parameters of the Gaussian functions (if Gaussians are used) of both distributions will be very close to each other in terms of the variable *length of the fish* (although, in terms of the multivariate distance they will not be very close overall, see Figure 5.12 line $|x_R^* x_{R+1}^*|$). The meaning for neural network is similar, but in terms of activation functions and their parameters.

As a tool to address this problem, Angelov *et al.* (2001) proposed to derive by simple averaging new parameters which are the same for both focal points that are close to each other in one (or few) dimensions:

$$x_{j^*} = \frac{x_{Rj^*} + x_{(R+1)j^*}}{2} \tag{5.16}$$

$$\sigma_{j^*} = \frac{\sigma_{Rj^*} + \sigma_{(R+1)j^*}}{2} \tag{5.17}$$

where

j^* denotes the particular input dimension in question (e.g. *length of the fish*);

for AnYa-type models Equation (5.17) is not necessary and Equation (5.16) transforms to $x_{j^*} \leftarrow x_{Rj^*}$.

This, effectively, leads to a simplification of the overall model in terms of a lower number of parameters and simpler linguistic interpretation (in the case of a fuzzy rule-based system) while preserving to a great extent the precision. This approach deals automatically with the issue of interpretability. In general, there are various approaches, especially in the area of fuzzy systems research, that, essentially introduce constraints of linguistic and interpretatability nature, which are to be taken into account by the system identification (in fact, optimisation) problems (de Oliveira, 1999). However, most of these approaches are suitable for offline applications only.

5.7 Conclusions

In this chapter the most important innovation represented by evolving systems structure is described. First, we describe the data space partitioning (including partitioning based on *prior* knowledge, regular partitioning methods, clustering, data clouds) and how this can be used to autonomously evolve the structure of the system from data streams in real time. In essence, following the millennia-old principle '*devide et impera*', complex problems can be decomposed into a set of (possibly overlapping) locally valid subsystems. Moreover, it has been demonstrated that this can be done online, in real time from streaming data. In the next section methods for autonomous monitoring the quality of the local submodels measured by their *utility*, *Age*, radius (in the case of clustering) are proposed.

Finally, methods for real-time, online normalisation and standardisation of the data samples are presented together with an analysis of the advantages and disadvantages of the alternatives. It should be stressed that these principles apply to various types of systems, including, but not limited to fuzzy rule-based, Bayesian, neural network, and so on. The models that are designed in this way are globally nonlinear yet locally can be treated as linear. When fuzzy rule-based systems are used, the models are also linguistically interpretable, which can be a big advantage for acceptability of these models by human operators. The approach that is adopted in this book, namely, evolving the system structure autonomously form data streams is fully unsupervised with regards to the system structure. Neither the number of submodels, nor their focal points or the effective number of inputs used needs to be predefined – instead it can be extracted from the data stream in an online mode and in real time.

6

Autonomous Learning Parameters of the Local Submodels

In the previous chapter the methods for structure design (if we use *prior* knowledge or regular data partitioning) and learning (if using clustering or data clouds) were described. It was mentioned that when using data clouds there are no parameters of the structure as such. In clustering, which are additional parameters (the position of the cluster centers which in some methods are means of the data samples, while in some other methods, such as *AutoCluster*, these are selected data samples as in the data *clouds*). These additional parameters concern the cluster radii.

Parameters that define the structure of the system are equal to the number of focal points that need to be selected (and possibly also cluster radii, if we use clustering instead of data clouds). It was described how to find them in the previous chapter. In this chapter the focus will be on the parameters of the local submodels (consequents). In general, local submodels can take various forms, for example singletons – zero order, linear – first order, Gaussian, triangular, trapezoid, polynomial, and so on.

Without loss of generality, we can assume linear submodels, because they include the simpler type of zero-order singletons as a special case and they are the most widely used type. For a set of locally valid linear submodels (see Figure 4.2) the task is to find the optimal values of parameters A in terms of minimising the error of prediction/classification/control/estimation/filtering. Before formalising and providing methods to solve this task in an online mode recursively the specifics of the particular problem will be stressed.

First, let us assume an evolving structure in which the number of local submodels is **not predetermined and fixed** and, thus, direct application of well-known and established techniques such as least squares (LS) (Hastie, Tibshirani and Friedman,

---

Autonomous Learning Systems: From Data Streams to Knowledge in Real-time, First Edition. Plamen Angelov.

2001) and, moreover, its recursive version, RLS (Astroem and Wittenmark, 1989) are, strongly speaking, not valid. In the next section, the offline LS method applied to multimodel systems will be briefly outlined. In addition, learning a complex multimodal system means that the criteria of optimisation can be defined either *globally*, in terms of the overall system or locally, per subsystem. The result of both types of optimality criteria will not be the same and each one has its own right to exist and its own meaning. This specific of the problem will be considered in more detail in Section 6.2.

6.1 Learning Parameters of Local Submodels

The overall output of the mutlimodel system can be given in a vector form as follows:

$$y = \psi^T A \tag{6.1}$$

where

$A = [a_1^T, a_2^T, \ldots, a_R^T]^T$ is a vector formed by the local (linear) submodel parameters;
$\psi = [\lambda_1 X^T, \ldots, \lambda_R X^T]^T$ is a vector of the inputs that are weighted by the normalised activation levels of the local submodels,
λ_i, i=[1,R] (if the model is zero order $\psi = [\lambda_1, \lambda_2, \ldots, \lambda_R]^T$); $X_k = [1,\ x_k^T]^T$.

Assuming the structure of the system is determined (see the previous chapter), the estimation of the local (linear) submodels parameters transforms into a LS problem.

For a given set of input–output data pairs (x_i^T, y_i), $i = [1,k]$, the (*global*) objective function is defined as (Angelov and Filev, 2003):

$$E^G = \sum_{i=1}^{k} \left(y_i - \psi_i^T A\right)^2 \rightarrow \min \tag{6.2}$$

where $\Psi_i = [\lambda_1(x_i)\, X_i^T, \lambda_2(x_i)\, X_i^T, \ldots, \lambda_R(x_i)\, X_i^T]^T$.

This objective function can be written in a vector form as (Angelov and Filev, 2003):

$$E^G = \left(Y - \Psi^T A\right)^T \left(Y - \Psi^T A\right) \rightarrow \min \tag{6.3}$$

where the matrix Ψ and vector Y are formed by ψ_i^T, and y_i, $k = [1, N]$, respectively.

Then, the vector Λ minimising Equation (6.3) could be obtained by the pseudo-inversion (Hastie, Tibshirani and Friedman, 2001):

$$A = (\Psi^T \Psi)^{-1} \Psi^T Y \tag{6.4}$$

The vector of (linear) local submodel parameters a minimising Equation (6.3) can more effectively be estimated online using the RLS algorithm (which itself is a simplified version of the Kalman filter):

$$\hat{a}_k = \hat{a}_{k-1} + C_k \psi_k \left(y_k - \psi_k^T \hat{a}_{k-1} \right) \tag{6.5}$$

$$C_k = C_{k-1} - \frac{C_{k-1} \psi_k \psi_k^T C_{k-1}}{1 + \psi_k^T C_{k-1} \psi_k} \tag{6.6}$$

with initial conditions and $C_0 = \Omega I$
where

Ω is a large positive number;
C is a $R(n+1) \times R(n+1)$ covariance matrix;
$\hat{a}_k$ is an estimation of the parameters based on k data samples.

6.2 Global versus Local Learning

The objective function, Equations (6.2) and (6.3), is *globally* optimal which guarantees best *overall* (*global*) performance in the whole data space on average (errors are summed). However, it does not guarantee locally optimal or even adequate behaviour of the local submodels, (see Figure 6.1 for a comparison). One can appreciate that locally optimal submodels do preserve the meaning approximating the data locally while the globally optimal submodels may achieve a better approximation overall, but this is for the expense of local interpretability (Angelov, Zhou and Klawonn, 2001; Yen and Gillespie, 2002).

In order to find *locally* meaningful submodels an objective function has to be locally weighted (Angelov and Filev, 2003):

$$E^L = \sum_{i=1}^{R} (Y - X^T a_i)^T \Lambda_i (Y - X^T a_i) \tag{6.7}$$

where

matrix X is formed by X_k^T; $\Xi \in R^{R(n+1)}$;
matrix Λ_I is a diagonal matrix with $\lambda_i(x_k)$ as its elements in the main diagonal (note that the weights λ depend on the input variables, x).

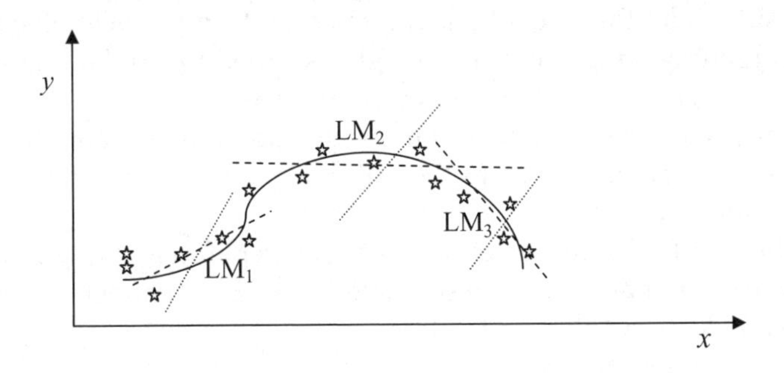

Figure 6.1 A simple 2D case of global (dotted lines) versus locally optimal submodels (dashed lines); the data are shown with stars

An approximate solution minimising the cost function (6.7) can be obtained by assuming that the linear subsystems are *loosely* (fuzzily) coupled with levels of interaction expressed by the weights $\lambda_i(x_k)$. Then the total cost, Equation (6.7) can be represented as a sum of local cost functions (Angelov and Filev, 2004):

$$E^L = \sum_{i=1}^{R} E_i^L \tag{6.8}$$

where $E_i^L = (Y - X^T a_i)^T \Lambda_i (Y - X^T a_i)$.

Solutions a_i that minimise the weighted LS problems expressed by the local error functions E_i^L can be obtained by applying a weighted pseudoinversion (Bishop, 2009):

$$a_i = \left(X^T \Lambda_i X\right)^{-1} X^T \Lambda_i Y \quad i = [1, R] \tag{6.9}$$

Alternatively, a set of solutions to individual cost functions E_i^L (vectors a_is) can be recursively calculated through the weighted RLS (wRLS) algorithm (Angelov and Filev, 2004). In this case, a wRLS algorithm that minimises each of the cost functions E_i^L is applied to the linear subsystem:

$$\hat{a}_{ik} = \hat{a}_{i(k-1)} + c_{ik}\, X_k\, \lambda_i(x_k)\, \left(y_k - X_k^T \hat{a}_{i(k-1)}\right) \tag{6.10}$$

$$c_{ik} = c_{ik-1} - \frac{\lambda_i(x_k)\, c_{ik-1} X_k X_k^T c_{ik-1}}{1 + \lambda_i(x_k)\, X_k^T c_{i(k-1)} X_k} \tag{6.11}$$

for $k = [1,R]$ and initial conditions $\hat{a}_0 = 0$ and $c_{i0} = \Omega I;\, a = [a_0, a_1, \ldots, a_n]^T$.

Note, that the covariance of the local submodel, c is not the same as the global covariance matrix, C.

The wRLS looks like the RLS with exponential forgetting (which is also, usually, denoted by λ), but the meaning is quite different. In wRLS λ represents the weights of each local submodel that are all taken into account in the overall system model. In RLS with forgetting λ represents the weight assigned to old data samples and the model is one, not a composition of simpler submodels.

In an extreme, when the weight, λ of a certain submodel is equal to 1 (obviously, the rest of the weights have to be 0) the wRLS algorithm reduces to a simple RLS based on this submodel.

When applying RLS locally to each submodel separately, Equations (6.10) and (6.11), the covariance matrices, c are separate for each submodel and have smaller dimensions ($c_{ik} \in R^{(n+1)\times(n+1)};\ i = [1, R]$).

The *locally weighted* RLS is significantly less affected by the structure evolution disturbance of the theoretical optimality for the RLS condition in comparison to the *globally weighted* RLS. In addition, it is computationally significantly less complex.

6.3 Evolving Systems Structure Recursively

In the *online* mode the system output is determined recursively by:

$$\hat{y}_k = \psi_k^T \hat{a}_{k-1} \quad k = 2, 3, \ldots \tag{6.12}$$

The '*hat*' means that these values are estimates as opposed to actual measurements. RLS is based on the assumption of a fixed system structure (a single linear model). It can be extended to the case of a mixture Gaussian or multiple other simpler models the number of which, however, is **fixed**. The optimisation problems (6.2) and (6.3) and, respectively, (6.7) and (6.8) are linear in parameters **under these assumptions**.

In an evolving scenario, however, the system structure is allowed to *gradually evolve* and, as a consequence, the number of submodels can vary, though the changes are, normally, quite rare (e.g. in a process with hundreds or thousands of time steps only a few dozen changes of the structure are, usually, observed). As a result of this evolution, the normalised firing strengths, λ_i can change. Since this evolution affects *all* the data (including the data collected before the moment of the change) the straightforward application of the RLS is not correct. A resetting is required of the RLS as described in (Angelov and Filev, 2004). This includes resetting the covariance matrices and initialisation of the submodels' parameters each time a new submodel is added or removed.

In the evolving structure scenario (when new subsystems are added or removed), the simplified Kalman filter is reset in the following way:

a. Parameters of the new submodel (in the time instant when a new submodel is added, that is $R \to R+1$) can be determined using the parameters of the existing submodels and taking a weighted average of them (Angelov and Filev, 2004), $\hat{a}_{R+1k} = \sum_{i=1}^{R} \lambda_i \hat{a}_{ik-1}$. The weights that are be used can be determined from the normalised activation firing levels of the existing submodels, λ_i; Covariance matrices are reset as

$$C_k = \begin{bmatrix} \rho\varsigma_{11} & \cdots & \rho\varsigma_{1R(n+1)} & 0 & \cdots & 0 \\ \cdots & \cdots & \cdots\cdots\cdots\cdots\cdots & \cdots & \cdots & \cdots \\ \rho\varsigma_{R(n+1)1} & \cdots & \rho\varsigma_{R(n+1)R(n+1)} & 0 & \cdots & 0 \\ 0 & 0 & 0 & \Omega & \cdots & 0 \\ \cdots & \cdots & \cdots & \cdots & \cdots & 0 \\ 0 & 0 & 0 & 0 & \cdots & \Omega \end{bmatrix} \tag{6.13}$$

where

ς_{ij} is an element of the covariance matrix ($i = [1, R \times (n+1)]$; $j = [1, R \times (n+1)]$);

$\rho = \frac{R^2+1}{R^2}$ is a coefficient.

b. Parameters of the other submodels can be *inherited* from the previous time step, $\hat{A}_k = \left[\hat{a}^T_{1(k-1)}, \hat{a}^T_{2(k-1)}, \ldots, \hat{a}^T_{R(k-1)}, \hat{a}^T_{(R+1)k}\right]^T$.
According to this approach, the covariance matrix is updated as follows:
   i. the last $n+1$ columns and last $n+1$ rows that are associated with the new, $(R+1)^{th}$ submodel are initialised with a large number, Ω in its main diagonal (as usual);
   ii. the remaining part of the covariance matrix (the top left $R \times R$ part) are multiplied by ρ.

This correction aims to compensate (or, rather, approximate) the effect it would have if changes (the new, $(R+1)^{th}$ submodel) had been in place form the very beginning. The derivation of the expression for ρ follows below (Angelov and Filev, 2004):

Starting from a vector of weighted inputs by the local densities (d_j) Ψ can be expressed as (Angelov and Filev, 2004):

$$\varphi_k = \left[d_1(x_k)\, X_k^T, d_2(x_k)\, X_k^T, \ldots, d_R(x_k)\, X_k^T\right]^T \tag{6.14}$$

$$\psi_k = \frac{1}{\sum_{j=1}^{R} d_j}\varphi_k \tag{6.15}$$

where d_j is the local density of the jth cloud/rule.

From the RLS (Equation (6.6)) the recursive update of the covariance matrix can be represented as:

$$C_k = C_{k-1} - \frac{C_{k-1}\varphi_k\varphi_k^T C_{k-1}}{\left(\sum_{j=1}^{R} d_j\right)^2 + \varphi_k^T C_{k-1}\varphi_k} \tag{6.16}$$

or expressing the history until the time instant **k** in an explicit way:

$$C_k = \Omega I - \sum_{i=1}^{k} \frac{B_i}{F_i + G} \tag{6.17}$$

where

$B_i = C_{i-1}\varphi_i\varphi_i^T C_{i-1}$;

$F_i = \varphi_i^T C_{i-1}\varphi_i$;

$G = \left(\sum_{j=1}^{R} d_j\right)^2$;

$C_0 = \Omega I$.

Let us suppose that the submodel added at the step k had been added from the very beginning (from time step 1) so, there are $R + 1$ submodels instead of R. Then, the covariance matrix at time k would have been:

$$\tilde{C}_k = \Omega I - \sum_{i=1}^{k} \frac{B_i}{F_i + \left(\sum_{j=1}^{R} d_j + d_{R+1}\right)^2} \tag{6.18}$$

or

$$\tilde{C}_k = \Omega I - \sum_{i=1}^{k} \frac{B_i}{F_i + G + \delta G_1 + \delta G_2}$$

where

$\delta G_1 = 2d_{R+1} \sum_{j=1}^{R} d_j;$

$\delta G_2 = d_{R+1}^2.$

It can be seen that adding a submodel at the time step k results in affecting the covariance matrix (Angelov and Filev, 2004), which is expressed in an increase of the denominator of the part subtracted from $C_0 = \Omega I$. Let us analyse the expression of the corrupted covariance. It can be seen that the values of δG_1 and δG_2 are strongly less than 1, B_i could be a big number since it is a quadratic form of the input data multiplied by the covariance matrix, F is bigger than δG_1 (since it is a sum of R positive values) while δG_1 is a single value only. F is also bigger than δG_2 if $d_{R+1} > \frac{1}{2}\sum_{j=1}^{R} d_j$.

Therefore, the role of the addends would be more significant only if *all* values of X_i (for *all* past time steps) or the covariance matrix tend to zero. The practical tests with a number of functions illustrate that the corruption of the covariance matrix by the addition of a new submodel is marginal and is absorbed by the remaining submodels.

This (small) influence can be approximated by a correction that amounts to an inverse mean. Indeed, if the corrupted covariance matrix is denoted by $\tilde{C}_{k+1}$ it can be expressed as some kind of function of the original one (C_{k+1}):

$$\tilde{C}_{k+1} = f(C_{k+1}) \tag{6.19}$$

Angelov and Filev (2004) proposed to use the inverse squared mean to express this function f:

$$\tilde{C}_{k+1} = C_{k+1}\frac{R^2+1}{R^2} = C_{k+1}\left(1 + \frac{1}{R^2}\right) = \rho C_{k+1} \tag{6.20}$$

From where we get $\rho = \left(1 + \frac{1}{R^2}\right)$.

In conclusion, the wRLS for the case of an evolving system structure (Angelov and Filev, 2003) is approximate, not exact and, thus, suboptimal.

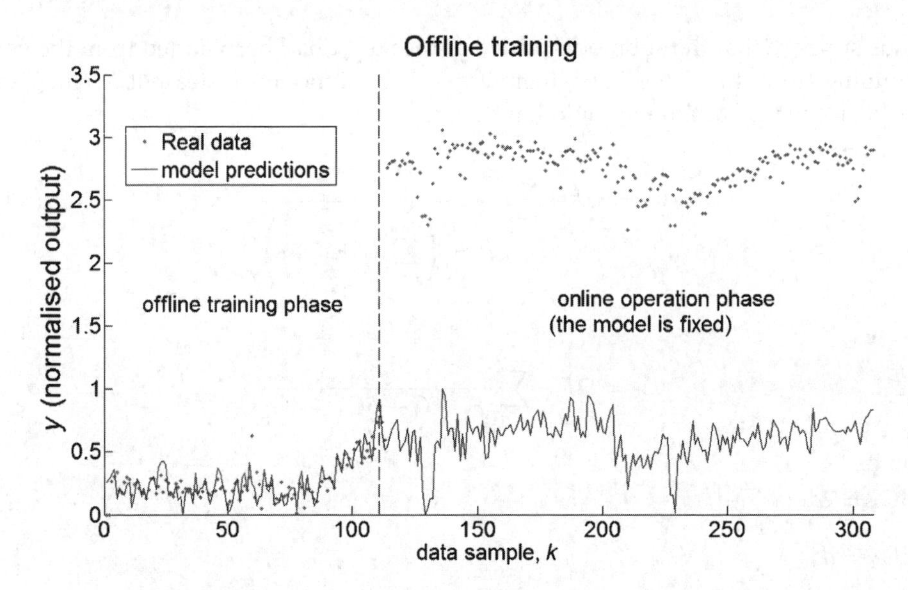

Figure 6.2 Offline training followed by online operation

6.4 Learning Modes

There are, generally, three modes of operation of an autonomous learning systems (ALS) that dependent on the level of process changes.

First, the traditional mode (Figure 6.2) includes an offline design phase followed by a clearly separated online operation when the model predicts reliably with the chosen *fixed* structure and parameters. Such models can be reliably used including in industry. For example, they formed the basis of the so-called self-calibrating, autonomous sensors (Angelov and Kordon, 2010) assuming that the process changes are negligible (for example, on average <5% outside the offline model development range).

For this particular example taken from the chemical industry (courtesy of Dr. Arthur Kordon, The Dow Chemical, USA) the predictions drastically deteriorate at sample 113 due to the change in the operating conditions and because the sensor (model) structure is *fixed* and does not have ability to evolve and to reflect the change in the data pattern. As a result, a significant error is generated.

A possible solution would be to collect enough data and retrain the original system model (the inferential sensor) in exactly the same way as the originally designed one (in an offline mode). The downsides of such an approach are the considerably increased cost of the development and maintenance and the overall lifecycle costs as well as the time of redevelopment and recalibration that is, usually, significantly larger than the sampling interval time. In addition, such a mode of operation leads to a complete loss of previously collected information and valuable knowledge. This second scenario may be required if the process changes are more significant (for example, 5 to 20% of the average deviation from the offline model development range).

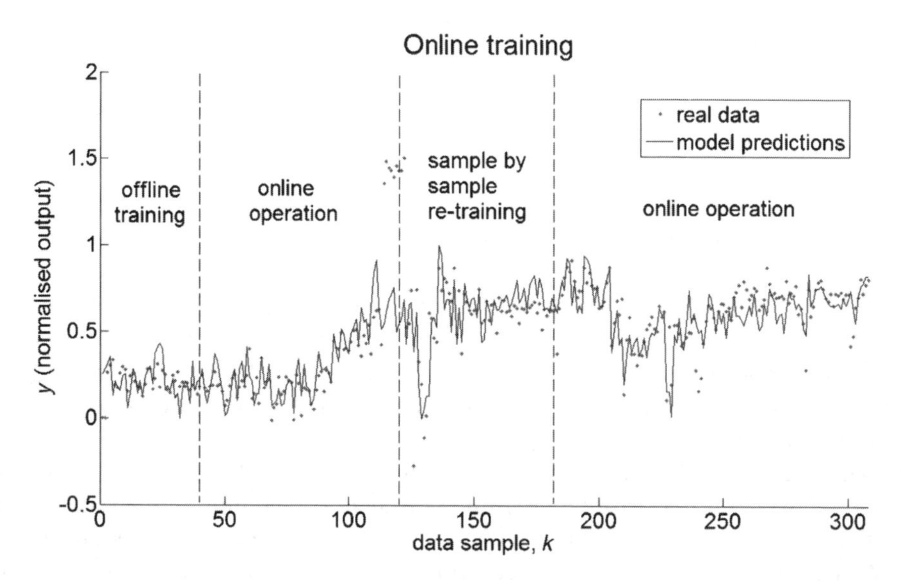

Figure 6.3 Evolving model structure prompted by the change in data pattern (around sample 113)

In Figure 6.3 the second submode is illustrated. The first submode would have its evolving (online training) phase after each new data sample immediately followed by a prediction or estimation or filtering (both model structure evolution and prediction/ estimation are performed before the next data sample arrives in a manner similar to the adaptive control (Astroem and Wittenmark, 1989) and online estimation (Widrow and Stearns, 1985; Bar-Shalom, Rong and Kirubarajan, 2001).

Finally, a third alternative is to continuously monitor and adapt/evolve the system structure to cope with and represent the significant changes in the data stream pattern. For example, as seen from Figures 6.2 and 6.3, a significant change of the data stream pattern takes place at sample 113 and the system structure adapts/evolves accordingly by adding a new subsystem online in order to react to this shift without interrupting the process of prediction. At the same time, this change is not replacing completely the previous system structure but keeps most of it (except one subsystem), thus, making the evolution process gradual and not abrupt (based on complete offline retraining).

When the data stream has frequent changes of its statistical characteristics (significant nonstationarity) the second mode (offline full retraining) has to take place quite often. If, in addition, the stream includes a lot of data, this becomes prohibitively costly to be done. In such cases the third mode of operation (evolving system structure) becomes necessary. The continuous adaptation of the local submodel parameters alone without the update of the structure cannot cope with these more significant changes.

In practice, the data samples needed for training usually come from measurements and might be collected over several sampling time intervals, which for different industrial processes may take seconds, minutes, hours, or even days (as is the case with oil refinery or waste water treatment processes) for example. The training samples may also come in batches. Therefore, the third mode of operation itself can be performed in the following two submodes:

a. evolve the system in online mode after each new data sample is available;
b. periodically evolve the system when new data is available.

In the latter case, during the periods of time when training data is not available, the inferential sensor will make predictions or estimations of the output based on the existing rule-base at the time. Obviously, when the feedback is available at each sampling time, the model of the inferential sensor can gradually evolve with each new data available.

Ideally, the system (e.g. an inferential sensor) should be able to adapt and retrain 'on the fly', without interrupting the online prediction each time a new data is available that can be used for training. This ability is critical when the system (e.g. an inferential sensor) is installed on a nonstop industrial system, which does not allow offline retraining like most of the industrial installations in the chemical, bio-, and petrochemical industries.

6.5 Robustness to Outliers in Autonomous Learning

If a data sample is identified to be an outlier (see Section 2.6) it should not influence the learning process, because this may introduce a bias. Therefore, once a data sample is identified to be an outlier using the RDE approach the local submodel parameters' learning is skipped for this time step.

6.6 Conclusions

In this chapter methods and algorithms for autonomous learning of the parameters of evolving systems are introduced. Stepping on the decomposed structure as described in Chapter 5 the task of learning parameters of local submodels seems significantly simplified. Indeed, local submodels are often linear (first order), singletons (zero order) or Gaussian, for which well-established offline learning techniques exist. However, the complexity of the problem is related to the fact that the system structure is **not fixed**, but **evolves**. Therefore, a specific, fuzzily weighted version of the well-known RLS is presented.

The optimality criteria (which for the case of a single and fixed linear model is linear in parameters) has different meaning in the *local* and *global* sense. This difference was thoroughly explained.

System structure (and, respectively, parameters) can be evolved in different modes of operation that are also described and illustrated with examples from a polymerisation process.

Finally, the problem with outliers in autonomous learning is addressed. In brief, a new data sample that is not described well by the existing system structure may be an outlier, but it also may be a seed of a new regime of operation or a new cluster or a data cloud. A distinction has to be made and a mechanism has to be developed to make the autonomous learning process more robust to such innovations. After all, the evolution is an innovation of the system structure, but not all anomalies are innovations, same as in the Nature!

It must be stressed that learning parameters of the submodels is a semisupervised problem because the true (actual/correct) values of the system output are assumed to be available to the system (at least during the online update periods).

7

Autonomous Predictors, Estimators, Filters, Inferential Sensors

The autonomous learning systems (ALS) concept described in this book is quite generic and can be applied to numerous problems. They can be summarised as:

A. clustering (unsupervised learning, multiple inputs, no output, MINO);
B. predictors, estimators, filters, inferential sensors (semisupervised learning, multiple inputs, multiple outputs, MIMO);
C. classifiers (semisupervised learning or unsupervised learning, MISO for the so-called two-class problem and MIMO for the general multiclass classification problem);
D. controllers (semisupervised learning; usually MISO, but can be MIMO).

Clustering was described in Section 3.2. In the context of ALS one can use *AutoCluster* or the ELM approach; that is, evolving clustering methods, which were described in Section 3.2.3.

7.1 Predictors, Estimators, Filters – Problem Formulation

In this chapter the problem B as itemised above will be described, namely, predictors, estimators, filters and inferential sensors. These seemingly different problems that are subject to various disciplines such as forecasting and statistical learning, signal

Autonomous Learning Systems: From Data Streams to Knowledge in Real-time, First Edition. Plamen Angelov.

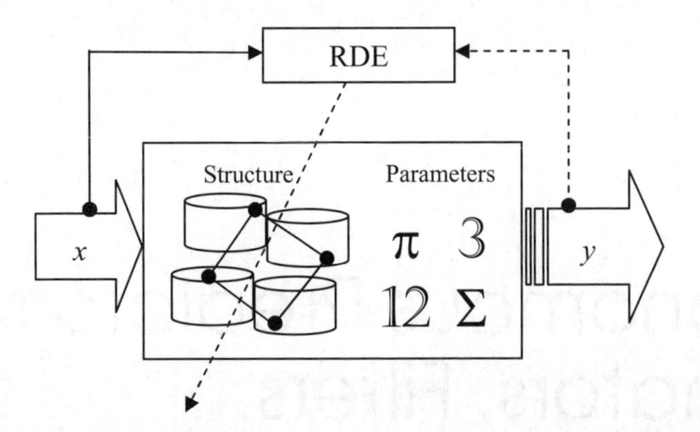

Figure 7.1 A schematic representation of the prediction, estimation, filtration and inferential sensors by an ALS. The feedback from the outputs (the true/target values are only taken when available, not necessarily after each sample – therefore, the link is shown with a dashed line); the adaptive feedback from the recursively estimated density concerns both the system structure and parameters

processing, chemical industry automation, system identification, and so on can be combined in the simplistic representation of Figure 7.1.

This extremely simplistic diagram represents a vector of inputs, $x \in R^n$ being transformed into the outputs, $y \in R^m$. The ALS has, broadly, a structure and parameters.

The difference between the filtering, estimation and prediction is best explained in terms of a time line, Figure 7.2.

The difference between estimation and filtering is illustrated in Figures 7.3 and 7.4.

From Figure 7.4 it is clear that the estimator provides a transformed value (an estimate) of the current, k^{th} value and the predictor provides the next/future, $(k + \Delta)^{\text{th}}$ value. In general, the step of prediction/filtering can be different from 1, as in this simple example. Apart from this important difference, all three transformations are quite similar.

Inferential (also known as *soft* or *intelligent*) sensors (Fortuna *et al.*, 2007) are algorithms or devices that are used to estimate or predict physical, chemical, biochemical and so on variables that are difficult to measure directly (by so-called *hard*/traditional sensors). Inferential sensors can be seen as a special case of estimators (if the current

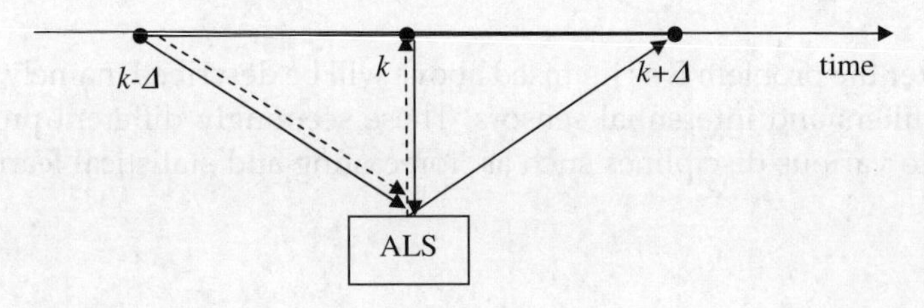

Figure 7.2 The difference between the prediction (solid line), estimation and filtration (dashed line), in terms of the time line. Δ denotes an integer number of time step

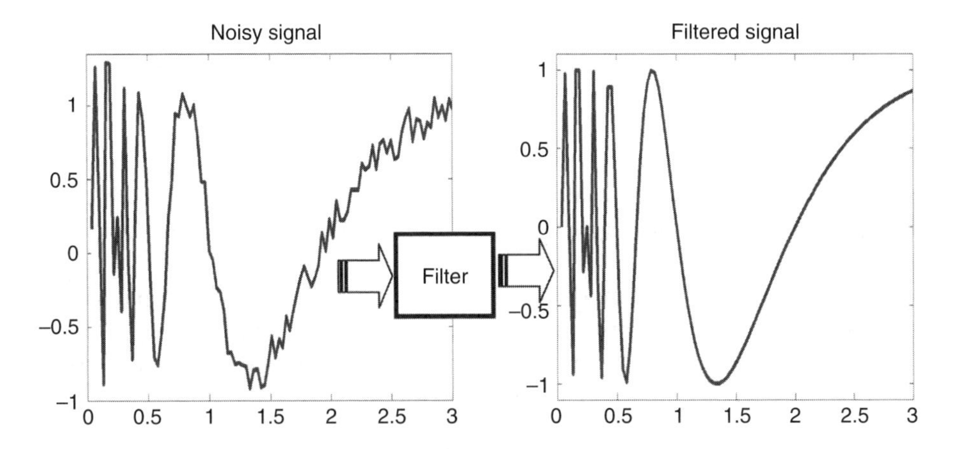

Figure 7.3 An illustration of the problem of filtration

value of the variable of interest is produced by the autonomous learning sensor) or predictors (if the next/future value is produced). This topic will be considered in more detail in Section 7.4.

7.2 Nonlinear Regression

Regression is a well-known problem from the statistical learning (Bishop, 2009; Hastie, Tibshirani and Friedman, 2001) that boils down to modelling (expression) of an output vector, y as a function of a vector of inputs, x:

$$y_k = f(x_k) \tag{7.1}$$

where f is a regression function performed, for example, by an ALS.

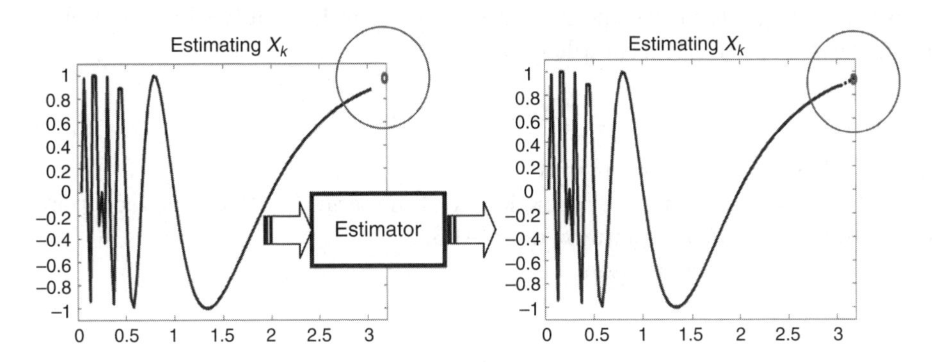

Figure 7.4 An illustration of the estimation problem

In the case of a regression, inputs and outputs have different physical (chemical, biochemical, social, etc.) meaning. In the case of a time series, both inputs and outputs represent the same physical (chemical, biological, social, etc.) variables, but differ by the time indices.

For example, the output can be '*the quality of the product produced*' or the so-called '*inflamability index*' in an oil refining process (Horak, 1993; Hernandez and Angelov, 2010) while the input vector can comprise density of the crude, pressure, temperature of the naphtha, and so on.

A specific of the regression models is that the input and output represent different physical variables. Usually (but not necessarily) they perform in estimation mode (that means, the output is made in the same moment of time as the inputs arrive, see Figure 7.2). They can also work in a prediction mode:

$$y_{k+\Delta} = f(x_k) \tag{7.2}$$

as well as in a filtering mode:

$$y_k = f(x_k) \tag{7.3}$$

The role of ALS is to represent the nonlinear regression, especially in case of a non-stationary data stream (data sequence with time-varying statistical characteristics). The nonlinearity and nonstationarity can be addressed by decomposing the original problem into a loosely (possibly, overlapping, fuzzy) combined locally (in the data space) valid simpler (usually, linear, singletons or Gaussian) regression models.

7.3 Time Series

Another popular problem (apart from regression) is the time series. This problem is also widely used in forecasting, statistical learning (Hastie, Tibshirani and Friedman, 2001) and signal processing (Haykin, 2002). In brief, it can be represented with a reference to Figures 7.1 and 7.2. A time series is a sequence of data samples. The main difference from the regression models is that both, the input, x_k and output vectors, $y = x_{k+\Delta}$ have the same physical meaning and the only differentiator is the time instances when a data sample is taken:

$$x_{k+\Delta} = f(x_k, A_k) \tag{7.4}$$

where A_k denote a parameter matrix.

Equation (7.4) represents the prediction of a time series (perhaps, the most popular, but not the only mode of operation of an ALS).

One can even have estimation:

$$\hat{x} = f(x_k, A_k) \tag{7.5}$$

where $\hat{x}$ is the estimated value.

Any problem of type B (see the beginning of the chapter) including regression as well as time-series analysis (including also inferential sensors that will be described in the next section) consists of two subproblems that are performed in real time for an interval of time shorter than the sampling time:

a. system structure identification; and
b. parameter estimation.

The first subproblem can be fully unsupervised (e.g. clustering, forming data clouds; it can also be regular partitioning or partitioning based on *prior*/expert knowledge).

As a result, focal points are generated that are representative, not very close to each other (overall and in terms of any specific input dimension) and cover well the whole data space. In addition, if not using AnYa method a parameter per focal point (cluster radii) is also determined as described in Chapter 5. If we use probabilistic models, these focal points are centers and the radii are spreads of Gaussian distributions.

If we use NFS this corresponds to layers 1 to 3 of the network (see Figure 4.6). If we use FRB systems this corresponds to the antecedent part of the rules:

$$Rule_i : IF(x_1 \sim x_{1i}^*)\, AND \ldots AND(x_n \sim x_{ni}^*) \tag{7.6}$$

The second subproblem is a semisupervised learning problem, as described in Chapter 6. It corresponds to adjusting the parameters of the local submodels (usually lines or singletons). If we use NFS it corresponds to layers 4 and 5, see Figure 4.6. If we consider a FRB system it corresponds to the consequent part.

7.4 Autonomous Learning Sensors

7.4.1 Autonomous Sensors – Problem Definition

Inferential or soft sensors can be seen as a special case of the systems of type B (see the beginning of the chapter for the description of the types) that have relatively wide application in various industries such as chemical and petrochemical, food and pharmaceuticals, automotive and other process industries (Liu, 2005). Their aim is to estimate or predict some important process variables that are difficult to measure directly representing them as mathematical functions of other available (easier to measure by traditional, also-called hard, sensors) variables.

In the engineering practice the true values of the important variables of interest (outputs of soft/intelligent/inferential sensors) are measured infrequently using laboratory analysis. This is necessary for calibration of the soft/intelligent/inferential sensors (to generate the target/ground truth values for the supervised learning). Laboratory analysis itself is a tedious and expensive offline process that usually requires interrupting the industrial process, which is very costly. It often includes material property tests, expensive gas chromatograph analysis, and so on.

The process monitoring and control using the target values derived offline can itself be online because the inputs are usually available online form cheap hardware sensors. It is, however, an open-loop process, because the structure of the soft/intelligent/inferential sensor is assumed once and is fixed (no longer changed). In practice, recalibration is usually done very rarely (e.g. once or twice per annum) because of the related costs (human involvement, interruption of the industrial process, computational costs, amount of new data). This adds to the overall lifecycle costs of the soft/intelligent/inferential sensors, making them less attractive to the industry users (Angelov and Kordon, 2010).

Soft/intelligent/inferential sensors can use various inference mechanisms. One can derive the output from so-called *first principles* if there is a clear understanding of the nature of the process. In this case one can estimate the parameters of the soft/intelligent/inferential sensor using an extended (nonlinear version of) Kalman filter (Kalman, 1960), extended kalman filters (EKF) (Bar-Shalom, Rong and Kirubarajan, 2001). In simpler cases the input-output relation can be approximated by a linear function and we use linear (multivariate) regression.

7.4.2 A Brief Overview of Soft/Intelligent/Inferential Sensors

Soft/intelligent/inferential sensors are widely used in industry due to their ability to provide accurate real-time estimates or predictions of difficult to measure variables of interest and replaces expensive measurements. Examples include product quality, inflamability index, emissions, biomass concentration, melt index, and so on that are being estimated or predicted from the cheap and widely available traditional/hard sensors like temperature, pressure, mass flow, and so on.

To maintain the quality of the products in the chemical and process industry, in general, it is a routine practice to take samples from the product during the processing (fermentation, rectification, etc.) due to the lack of or difficulties related to the measurement of some parameters such as concentrations, product quality, and so on. Samples are usually taken with intervals of a few hours and analysed in a laboratory environment. The main aim is to certify the process by monitoring the deviation from a specification. Another scenario includes modelling at the design and process planning stage – the inference between certain measurable process variables and certain target value (product quality, for example) is of great importance.

Many of the *soft/intelligent/*inferential sensors available on the market and used in industries (provided by vendors, such as Pavilion Technologies, Aspen Technologies, Honeywell, Siemens, Gensym, etc.) use black-box type of models such as neural network, SVM, genetic programming, statistical and empirical models. Such models, however, have high lifecycle cost because of the high development as well as maintenance costs related to manual interventions and computational expertise required.

In reality, the environment in which the industrial process and *soft/intelligent/* inferential sensors operate is dynamically changing, evolving; the equipment is wearing and contaminated, often being replaced. Any (even minute) process changes outside the conditions which were used for the offline *soft/intelligent*/inferential sensor development can lead to performance deterioration that requires maintenance and recalibration.

This requires redesign of the sensor/model, including derivation of an entirely new structure. In this case, modelling expertise is needed and, as a result, maintenance costs are increased

In addition, a number of aspects of the industrial processes are often ignored in practice due to their complexity; raw materials (which are represented by the input variables) alter in quality. The main weakness of such sensors is that significant efforts are involved in their development (which is offline, based on a batch set of data) and maintenance (which includes laboratory tests and process interruptions).

7.4.3 Autonomous Intelligent Sensors (AutoSense)

An alternative was proposed recently by Macias *et al.*(2006) and Angelov *et al.* (2008) in the form of evolving (self-calibrating, self-maintaining, autonomous) inferential/*soft/intelligent* sensors that are flexible to the extent that they can adapt their structure as well as parameters in order to follow the data pattern, to retrain, and recalibrate 'on the fly'. They were reported (Angelov and Kordon, 2010; Ferreyra and Rubio, 2006) to save time, computational and human resources. These autonomous sensors have an additional important advantage in that they are also transparent and interpretable, because they use FRB models.

The gradual evolution means that the model structure remains largely the same with only one fuzzy rule (or neuron or Gaussian local pdf) being added or removed from time to time (not very often – only when a significant change in the data density pattern takes place, see Figure 6.3).

Soft/intelligent/inferential sensors are, in general, nonlinear. This is related to challenges, such as unpredictable extrapolation, lack of model confidence limits, and multiplicity of model solutions. The reliable performance with acceptable accuracy of prediction/estimation inside the range of the training data cannot be guaranteed because the data may (and usually are, in practice) nonstationary. Since the process and operating condition changes are rather a rule than an exception, sensor recalibration or even complete redesign is often required which increases the lifecycle costs significantly (Angelov and Kordon, 2010).

The main reasons for the economic benefits that *soft/intelligent*/inferential sensors provide can be summarised as follows (Hernandez and Angelov, 2010):

- they allow tighter control of the most critical parameters of final product quality and, as a result, the product consistency is significantly improved;
- they reduce upsets of the industrial processes due to early detection of possible problems via online estimation of critical variables;
- they improve working conditions by reducing or eliminating laboratory measurements in dangerous environments;

- they are very often optimal from the economical point of view;
- their development and maintenance cost is lower in comparison to the alternative solutions (hardware or *first principles* models);
- they reduce capital investments by optimising the use of expensive hardware;
- they can be used not only for estimation and prediction, but also for running "*what-if*" scenarios in production planning.

Due to these economic benefits the process industries started widely to successfully develop and deploy *soft/intelligent*/inferential sensors during the last twenty years.

The most popular application area of *soft/intelligent*/inferential sensors is the environmental emission monitoring (Qin, Yue and Dunia, 1997). For example, NO_x emissions in burners, heaters, incinerators, and so on, can be inferred from associated process variables, like temperatures, pressures, and flows. Traditionally, emission monitoring is performed by expensive analytical instruments with costs in the order of hundreds of thousands of pounds plus maintenance costs in the order of tens of thousands of pounds per annum. The *soft/intelligent*/inferential sensors offer a much cheaper alternative with acceptable accuracy (Kordon, 2006). Therefore, since the mid-1990s according to the leading vendor in *soft* sensors for emission monitoring Pavilion Technologies, over 250 predictive emission monitoring, PEMs *soft* sensors have been installed.

Another area where *soft/intelligent*/inferential sensors were successfully applied is the estimation of biomass (also-called cell mass) concentration in continuous and fed-batch bioprocesses (mainly fermentations) used in food, pharmaceutical and other industries (Chen *et al.*, 2004). Estimating the cell mass concentration is pivotal for the control of fermentation processes, especially during the micro- organisms' growth phase (Angelov and Tzonkov, 1993). Traditionally, the cell mass concentration is determined offline by a laboratory analysis every 2 to 4 hours. This low frequency leads to a poor control performance that can partially be compensated by online estimates. This is the rationale for the use of *soft/intelligent*/inferential sensors for this problem.

Yet another popular application of *soft/intelligent*/inferential sensors is the estimation of product composition in distillation columns and prediction of polymer quality in terms of the melt index, average molecular weight, polymerisation rate or conversion that are inferred from the reactor's, jacket inlet and outlet temperatures, and the coolant flow rate through the jacket (Kordon *et al.*, 2003). *Soft/intelligent*/inferential sensors can also be used to estimate online the amount of reactor impurities during the initial stage of the polymerisation (Kordon *et al.*, 2003).

7.4.4 AutoSense Architecture

The architecture of autonomous learning sensors *AutoSense* can be represented as in Figure 7.5.

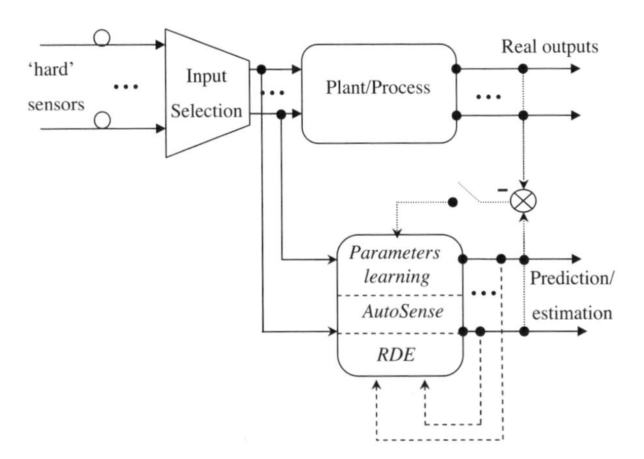

Figure 7.5 Architecture of *AutoSense*. The dashed lines are active only from time to time when true information is available (e.g. by laboratory tests or other) and is used for automatic supervisory retraining. They indicate the use of the predicted outputs in (recursive) density estimation based on which the sensor/model structure may need to evolve

The procedure of *AutoSense* can be represented by a pseudocode as shown in Appendix B5. The main advantages of *AutoSense* are that:

- it is evolving, self-calibrating and self-maintaining (they are not pretrained and fixed);
- learning can start 'from scratch' with the very first data sample or from *a priori* information (if available);
- it can have a MIMO structure and, thus, build a separate regression model for each output variable;
- it can ensure high prediction rates;
- it is one-pass, recursive and have low computational requirements, which makes it suitable for hardware (on chip) implementation;
- it contributes to the online analysis and monitoring of industrial processes including drift of the data density in the data stream, anomalies and possible faults.

7.4.5 Modes of Operation of AutoSense

Generally, inferential sensors have three modes of operation, as described in Section 6.4. The first mode is the online operation when they predict reliably with a *prefixed* structure and parameters.

The industrial experience from the chemical industry shows that the inferential/*soft sensors* in practical use demonstrate acceptable performance in this mode if the process changes are minor, that is, on average <5% outside the offline model development range. In the case of more significant process changes (i.e. on average 5–20% outside the offline model development range), the sensor starts to provide higher errors and needs recalibration or retraining. The most frequent solution is offline refitting the model parameters to the new process conditions. However, in the case of frequent changes, a continuous adaptation to the new conditions is needed through parameter estimation within the fixed structure.

The third mode of operation of inferential/*soft* sensors, handles the most difficult case of major process changes (i.e. on average >20% outside the offline model development range) when the sensor model changes automatically its structure and corresponding parameters to adjust to the new data density pattern.

7.4.6 Autonomous Input Variable Selection

An important part of the autonomous learning sensors design and development is the automatic input variables selection, as described in Section 5.4.1. The process of input selection is usually a part of the preprocessing and traditionally is performed either by the operator (human, model designer, decision maker) or by approaches that are offline such as PCA, PLS, independent component analysis (ICA), and so on.

The approach proposed and described in Section 5.4.1 offers a continuous monitoring of the modelling process and removing the inputs/features that do not contribute significantly to the output. This simplifies the overall system and affects the dimensionality of the covariation matrix in particular, of the vector of inputs used and the related parameters.

An example of the choice of the input variables to be used based on the weight of their contribution as described in Section 5.4.1 is depicted in Figure 7.6

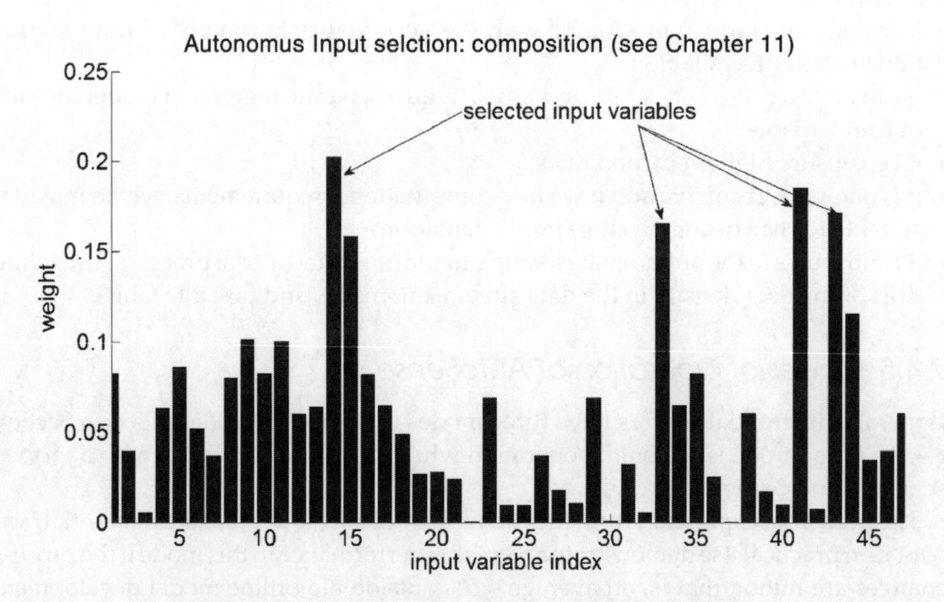

Figure 7.6 An example of input selection by an ALS

7.5 Conclusions

In this chapter several types of models were described that can be combined into a single one (denoted as type B in the classification made in the beginning of this chapter). These include predictors, estimators, filters as well as inferential (*soft/intelligent*) sensors that can be seen as a special case of the above.

These types of models are, generally MIMO; they take multiple inputs and produce (possibly) multiple outputs. The main difference is the position of the output with regards to the input in terms of time. They play an important role in signal processing, statistical analysis, econometrics and other disciplines and with the proposed ALS approach they can be evolved online real time in an autonomous manner even if they are nonlinear and nonstationary.

These types of problems can take a form of a regression or a time series. In addition, special attention is given to the particular type of such models in the form of so-called inferential (also-called commercially *soft* or *intelligent*) sensors. This specific type of predictors or estimators have significant commercial importance especially in chemical, petrochemical and related process industries and will be further described from the application point of view in Chapter 12.

8

Autonomous Learning Classifiers

8.1 Classifying Data Streams

One traditional approach to classifying data streams is the incremental classifier (Fung and Mangasariany, 2002). In the literature there are various classification frameworks that work in an incremental mode (per sample), for example, decision trees (Yuan and Shaw, 1995), neural network such as adaptive resonance theory, ART (Carpenter and Grossberg, 2003), incremental learning vector quantiser, iLVQ (Poirier and Ferrieux, 1991), probabilistic such as incremental versions of Bayesian classifiers (Schlimmer and Fisher, 1986), incremental Fisher LDA (Pang, Ozawa and Kasabov, 2004), and so on. It should be stressed, however, that the classifier structure in all incremental classifier methods mentioned above is ***fixed***.

Incremental classifiers are inefficient with respect to the problem of the so-called *drift* and *shift* in the data density pattern. In machine learning by *drift* they refer to a modification of the concept over time that relates to a relatively smooth transition of the data distribution from one local region of the feature space to another (Widmer and Kubat, 1996). It is author's point of view that *drift* and *shift* have to be considered from the point of view of data density (not pdf but the density as described earlier in this book).

By *shift* they traditionally refer in machine learning literature to a more abrupt change such as the sudden appearance of a fault or an abrupt change of a regime of operation (Tsymbal, 2004). In order to represent such *sudden* and *abrupt* changes tuning only (consequent) parameters of a classifier is not enough. In this respect the autonomously learning and self-evolving classifiers described in this chapter are particularly suitable, because they allow changing the classifier structure dynamically and automatically.

Autonomous Learning Systems: From Data Streams to Knowledge in Real-time, First Edition. Plamen Angelov.

8.2 Why Adapt the Classifier Structure?

In practice, nowadays, the classifiers need to cope with large quantities of data, often streaming with a fast rate. The challenges that classifiers face are related to the need to address

a. nonlinearity and nonstationarity;
b. large or even huge amounts of data;
c. real-time, recursive processing.

Most of the traditional classifiers (e.g. the one described in Section 4.4, for example) are designed to operate in batch (offline) mode and have a fixed structure (pdfs, fuzzy or decision rules, neurons, etc.). This fixed structure corresponds to a fixed classification hypersurface.

In reality, however, new data samples that arrive during the operation of the classifier do not necessarily follow the same distribution as the training data (the nonstationarity of which is well known to lead to problems such as overfitting, low generalisation and *drift* and *shift* of the density in the data stream (Lughofer, Angelov and Zhou, 2007)).

A more efficient alternative is to use adaptive, self-learning, also called evolving classifiers such as eClass (Angelov, Zhou and Klawonn, 2007; Angelov and Zhou, 2008), FLEXFIS-Class (Lughofer, 2011), AnYa-Class (Angelov and Yager, 2012). They are able to capture and react to the changes in the density evolution of the data pattern present in the data stream during the operation of the classifier and self-develop/evolve.

The *AutoClassify* algorithm presented here can be seen as a NFS or a FRB but same principles apply to probabilistic (Bayesian) classifiers also if consider local (per cluster) Gaussian pdfs, for example.

One example where such classifiers are very useful is the automatic classification of the behaviour of users (Iglesias *et al.*, 2009). Another strong example is the area of so-called cybersecurity related to network intrusion detection and classification (Angelov and Zhou, 2008; Baruah and Angelov, 2012). In both cases, if we assume a classifier with a prefixed structure it will be unable to distinguish a new type of behaviour (new type of users) or a new type of threat. At the same time, it is well known that users change their behaviour and a new type of users may join a service/office or leave it (there is a dynamic element that is often ignored and an element of evolution of the behaviour). Hackers are also well known to be inventive and try ever newer 'tricks'.

In this sense, a traditional offline trained and with a prefixed structure classifier will only be able to recognise behaviour of certain types of users and of certain attacks to the network, respectively, while an evolving and self-learning classifier will be able to autonomously evolve its structure to adapt to a potentially changing data pattern without expensive full retraining, redesign and human involvement.

In effect, the traditional, offline classifiers (Ishibuchi *et al.*, 2004) are valid only to a 'snapshot' of the data stream and require *all* the previous data for a possible retraining and redesign, which is costly from computation and memory usage point of view. In contrast, *AutoClassify* works on a per-sample basis and only requires the features of that sample plus a small amount of recursively updated information related to the density. In addition, it can also performs online feature selection, as described in Section 5.4.1. It does *not* require *the history* of *all* past data samples yet it takes into account the data distribution (through the RDE) of *all* past data *exactly*. *AutoClassify* is a *one-pass* (noniterative) algorithm – each sample is processed only once at a time and is then discarded from the memory.

8.3 Architecture of Autonomous Classifiers of the Family AutoClassify

A classifier is a mapping from the feature space to the class label space. An ALS of type C (as itemised at the beginning of Chapter 7) can be used to autonomously learn and adapt the classifier as shown in Figure 8.1.

The main differences (if comparing this scheme with the similar scheme from Figure 7.1, type B ALS) is that the outputs are class labels (usually integer values) and that a (recursive) preprocessing phase is added. Note, that all three phases are performed for a time shorter than the sampling period (between the current time instant, k and the next time instant, $k + 1$). Also, the feedback with the ground truth (target) labels is only available for *AutoClassify1* (see Section 8.2) and this is not necessarily the case for each data sample, but only when available.

As a consequence of the different form of the output vector, the inference of classification problems is also different – it is usually of the 'winner takes all' type,

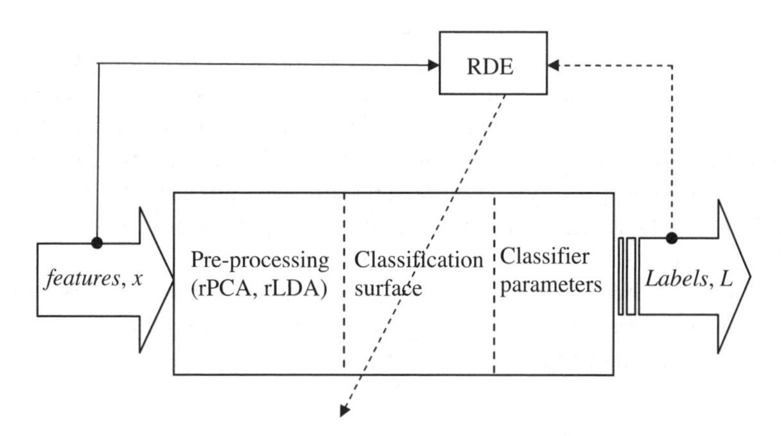

Figure 8.1 A general scheme of *AutoClassify*. $x \in R^n$ denotes the vector of features; $L \in R^m$ denotes the vector of labels

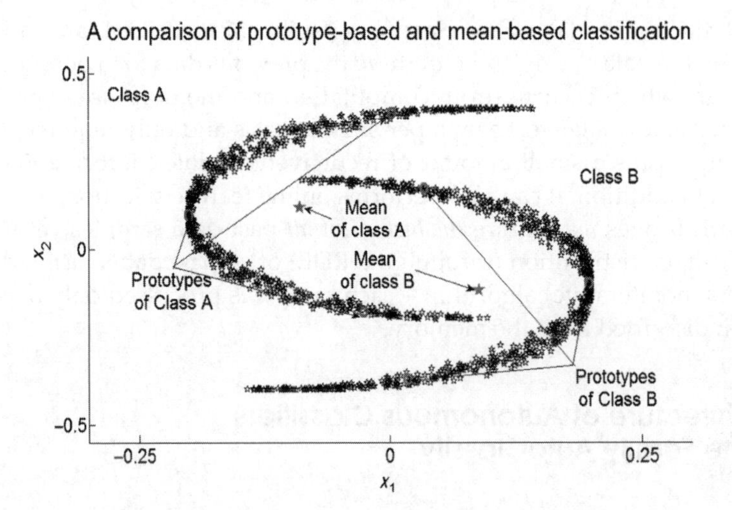

Figure 8.2 An illustration of the prototype-based nature of *AutoClassify*

Equation (4.21), although not always – it will be shown later that CoG type of inference, Equation (4.24), is also possible to be used.

The *AutoClassify* family consists of prototype-based classifiers. The well-known *k* nearest neighbours, *kNN* classifier (Duda, Hart and Stork, 2000) is also a prototype-based one. However, the majority of the other classifiers are mean based.

The difference between the two approaches can be illustrated by the following example. Let us have a 2D problem, as represented in Figure 8.2. If we apply a mean-based approach (such as *k* means, Bayesian or a traditional FRB classifier, LVQ or RBF neural network, etc.) the class label (*A* or *B*) will be determined by the distance between the new validation sample (previously unseen by the classifier) and the class means (stars of red and green colour, respectively).

It is obvious that if the new data point is close to many of the points of class *B*, which are close to the mean of class *A*, it will be misclassified. Finally, if it is close to many points of class *A* that are close to the mean of the class *A* it will be misclassified, too.

At the same time, the prototype-based approach has clustering (or data clouds) as its first stage. The prototype points have another advantage that they are always feasible points, while means are abstract, nonexisting points that may even be infeasible.

In addition, FRB or NFS classifier (and, to some extent, probabilistic and decision tree-type ones) unlike multilayer perceptron type neural network classifier have interpretable and human-intelligible structure.

8.3.1 AutoClassify0

The main difference between the zero order *AutoClassify0* and the first-order *AutoClassify1* is in the form of their consequent part. In *AutoClassify0* it is represented by the class labels directly as in Equation (4.19). The inference is produced using the 'winner takes all' principle, see Equation (4.21) and several rules per class are formed. The main difference from the traditional offline (e.g. FRB, NFS or Bayesian) classifiers is that the classifier structure (rules, neurons, pdfs) is not fixed, but evolves autonomously. The focal points around which the local submodels (singletons, linear ones, or Gaussian pdfs) are being later developed are determined based on the data density using *AutoCluster* or ELM or forming data *clouds* using AnYa method.

8.3.2 AutoClassify1

The architecture of *AutoClassify1* differs significantly from the architecture of *AutoClassify0* by the use of linear submodels in the consequent part, Equation (4.20). In the case of the probabilistic model the consequent part may be represented by locally valid Gaussian pdfs. *AutoClassify* takes into account the fact that the classification surface is dynamically changing/evolving and tries to approximate it by learning from the data stream.

The difference between *AutoClassify1* (ALS type C) and the ALS type B (regression and prediction models) is that *AutoClassify1* (ALS type C) has m groups of submodels (per class) and, respectively, the averaging is performed accordingly – the outputs of particular local submodels are normalised (to sum up to 1) by:

$$\overline{y}_i = \frac{y_i}{\sum_{j=1}^{R_l} y_j} \tag{8.1}$$

where

$$\sum_{i=1}^{R_l} \overline{y}_i = 1;$$

R_l denotes the number of local submodels per class $l = [1, m]$.

The overall output is formed as a weighted sum of the ***normalised*** outputs of each local submodel using CoG inference (unlike *AutoClassify0* that is using the 'winner takes all' inference):

$$y = \sum_{i=1}^{R} \frac{d_i}{\sum_{j=1}^{R} d_j} \overline{y}_i \tag{8.2}$$

If use FRB or NFS, *AutoClassify1* can also be interpreted as a combination of loosely (fuzzily) linked locally valid LDAs.

8.3.2.1 Multiple (m) Two-Class Classification Problems

In general, *AutoClassify1* considers a MIMO (multiclass) classifier, but the specific case of a so-called two-class classifier ($m = 2$) is quite popular and widespread. In many cases multiclass classification problems are considered as a combination of m two-class classification problems (Bishop, 2009). If that is the case then parameters of the local linear submodels, a_i form a vector-column (index i will be omitted for clarity) $\vec{a} = [a_{01} \ a_{11} \dots a_{n1}]^T$.

The output is used then to discriminate between the two classes. For example, using a threshold of 0.5 outputs above 0.5 are classified as one of the two classes, while outputs below 0.5 are classified as the other class:

$$\begin{aligned} &IF(y > 0.5) \\ &THEN(Class\ A) \\ &ELSE(Class\ B) \end{aligned} \tag{8.3}$$

8.3.2.2 *AutoClass1* MIMO

When an overall MIMO model is used, parameters form a $(n + 1) \times m$ matrix as in Equation (4.11). In *AutoClassify1* MIMO, the outputs form a m-dimensional vector row – one normalised output for each class, $\overline{y}_i = [\overline{y}_{i1}, \overline{y}_{i2}, \dots, \overline{y}_{im}]$. Each of the m '*wining*' labels are determined by the highest value of the normalised outputs per class, $\overline{y}_l$ (Angelov and Zhou, 2008):

$$L = L_i^*; \ i^* = \arg\max_{l=1}^{m} \overline{y}_l \tag{8.4}$$

Note, that Equation (8.4) is not the typical 'winner takes all' principle in terms of the firing strength of the rules as applied in classification problems. It resembles more the LDA and SVM rather than typical FRB classifiers.

The target labels are usually 0 and 1 (where 1 is used for the membership to a class) and, therefore, the normalised outputs per local submodel, $\overline{y}_{ij}$; $j = [1, m]$ can be interpreted as the *possibility* of a data sample belonging to the respective class. In this context, *AutoClassify1* MIMO resembles the so-called indicator matrix approach (Hastie, Tibshirani and Friedman, 2001) which is used for traditional offline classifiers.

It is interesting to note that *AutoClassify1* MIMO can also be used for a two-class classification problem. Then, the target outputs are two dimensional. For example, using the notations from Equation (8.3) then $y = [0 \quad 1]$ if the class label is B. The number of parameters (of the consequent part) in this case will, however, be twice as large:

$(2 \times (n + 1)$ since $\vec{a} = \begin{bmatrix} a_{01} & a_{11} & \dots & a_{n1} \\ a_{02} & a_{12} & \dots & a_{n2} \end{bmatrix}^T$ in comparison to *AutoClassify1 MISO* $n + 1$ parameters since $\vec{a} = [a_{01} \quad a_{11} \quad \dots \quad a_{n1}]^T$).

8.4 Learning *AutoClassify* from Streaming Data

8.4.1 Learning AutoClassify0

Learning *AutoClassify0* is unsupervised and is based on focal points by clustering or partitioning into data clouds. The class label of the focal points is then used for 'winner takes all' (or 'few winners take all', which is similar to the kNN approach) inference mechanism, similarly to Equation (8.4).

Learning of *AutoClassify0* is, therefore, largely described by the clustering method (or method for forming data clouds), which can be *AutoCust*, ELM, and so on. The joint features-labels data space is considered, $z = [x^T, L]^T$ *per class*. In this manner, a number of focal points (respectively, clusters or clouds) are formed for each class. Then, around the prototypes (focal points) either fuzzy rules or local Gaussians are being formed. This leads to the formation of information granules, which can be described linguistically and in this way the data is automatically transformed into primitive forms of knowledge.

It is important to stress that the overall model is composed of m subgroups of local submodels and in each subgroup the consequents (labels) of all submodels are the same, but the number of submodels, $R_l \geq 1$ $(l = 1, 2, \ldots, m)$ can be different. The total number of submodels, $R = R_1 + R_2 + R_m$ should be at least as big as the number of classes, m: $R \geq m$.

Every new data sample with a class label that has not been previously seen becomes automatically in real time a new prototype (focal point). However, this prototype is often later replaced by more descriptive prototypes, because there is a prototype replacement and removal mechanism according to the principle C (Section 5.2.5) as described in Section 5.4. Therefore, *AutoClassify0* can learn autonomously even starting 'from scratch' and without knowing the number of classes in advance! This is a unique property (especially for areas such as mobile robotics) as will be described in Chapter 12.

The procedure of *AutoClassify0* is summarised in the pseudo-code given in the Appendix B6.

8.4.2 Learning AutoClassify1

Learning of *AutoClassify1* is very similar to the ALS of type B, which usually also consists of first order (with linear consequents) submodels. Learning is based on the decomposition of the identification problem into:

a. overall system structure design; and
b. parameter identification.

Both of these subproblems can be performed in the online mode during a single time step (per sample) similarly to the way adaptive control and estimation work – in the period between two samples two phases are performed:

a. classification; and
b. classifier update (from time to time when necessary – see the principles A to C from Section 5.2.5).

During the first phase the class label is not known and is being predicted; during the second phase, however, it is known and is used as a supervisory information to update the classifier (including its ***structure evolution*** as well as its parameters update). In this sense, *AutoClassify1* is using a semisupervised learning method and the training samples with correct/true labels are required.

The algorithm procedure is given in Appendix B7.

8.5 Analysis of *AutoClassify*

The main novelty of *AutoClassify* can be summarised as follows:

- It is evolving (the classifier is not pretrained and fixed; learning can start 'from scratch' with the very first data sample);
- It can have a *MIMO* structure and, thus, build a separate regression model for each class. If a sample with a previously unseen class label is met the *MIMO* structure of *AutoClassify* expands naturally by initialising learning of the new $(R+1)^{th}$ class from this point onwards in the same way as for the remaining R classes;
- It can attain high classification rates comparing favourably with well-known offline and incremental classifiers;
- It is one-pass, recursive, and, therefore, has extremely low memory requirements (therefore, suitable for hardware including on-chip implementations);
- It is useful for online analysis and monitoring of the *drift* of the density in data streams using the concept of monitoring the *age* of the local submodels.

If we compare *AutoClassify0* with *AutoClassify1*, in general, the former stands out with its unsupervised nature, but its performance (classification rate) is usually significantly lower because the zero-order local submodels have a lower degree of freedom compared to the first-order *AutoClassify1*.

8.6 Conclusions

In this chapter the autonomous self-learning classifier family *AutoClassify* is presented. This is an ALS of type C (as described at the beginning of Chapter 7). The main differences if compare with the ALS of type B are that the outputs are class labels (usually integer values) and, therefore, this approach, although being able to work in an unsupervised manner, has a lower degree of freedom, which leads to significantly lower performance.

AutoClassify0 is very interesting and attractive by the fact that it is completely unsupervised and only at the level of label interpretation human involvement can take place, but not a must. The class labels can be assigned by default as sequential numbers, but more meaningful labels can also be provided.

AutoClassify1, the other alternative type of autonomous learning classifier, is of first order, which allows much better performance to be achieved, but is semisupervised and has more parameters (of the local submodels). *AutoClassify1* can work as a MIMO type of model for multiclass classification problems. It can also solve the same type of problems by applying m (where m is the number of classes) two-class classification problems (*AutoClassify1* MISO problem).

These types of classifiers with evolving, self-developing structure have been first introduced by Angelov, Zhou and Klawonn (2007), Angelov, Ramezani and Zhou (2008) some five years ago and here they are put in the context of ALS. The results of their application to the practical problems of landmark recognition, scene recognition, evolving user behaviour modelling and other related problems from mobile robotics, ambient assisted living and ubiquitous computing will be described in Chapters 12–14.

9

Autonomous Learning Controllers

The idea for self-learning controllers is not new and is, perhaps, at the origins of the very idea of self-learning and self-organising systems taking its roots from the very strong, at that time, Moscow Institute of Control Problems (IPU), (IPU was also the work place of Vladimir Vapnik, the 'father' of SUM) and mainly related to the works of Tsypkin (1968). This gave the seed for the powerful modern adaptive control theory (Ljung, 1987; Astroem and Wittenmark, 1989). It was and still is, however, mostly valid for linear systems (Kailath *et al.*, 2000) or so-called Hammerstein-type quadratic models and concerns parameter tuning rather than system-structure adaptation.

Later, Procyk and Mamdani (1979) proposed their self-organising fuzzy logic controller (FLC) that was, however, confined to a fixed-size look-up-table, thus, the structure adaptation was very limited and related to the choice of predefined fuzzy sets. Narendra and Parthasarathy (1990) extended the adaptive control systems theory to NN-based multimodel systems, but this was again limited to the case of a fixed system structure and concerned parameter tuning only. Psaltis, Sideris and Yamamura (1988) proposed to model the inverse plant dynamic in an adaptive control scheme using an offline-trained NN and use this to derive controller that would get the performance (output) as desired assuming the plant dynamic is perfectly modelled.

Angelov (2002, 2004b) and Angleov and Buswell (2001), proposed to use evolving FRB to model the inverse plant dynamic and coined the self-evolving FRB eController and the self-evolving *p*arameter-free *r*ule-based *c*ontroller, SPARC (Sadeghi-Tehran *et al.*, 2012) and OSECC (Angelov, Skrjanc and Blazic, 2012). eController combined the benefits of the FRB controllers, namely that they do not require the plant and environment (mathematical/physical) models to be known, that they can use and operate with linguistic and human-intelligible knowledge and are proven universal approximators (Wang, 1992) with the advantages of self-learning online and dynamically evolving structure (fuzzy linguistic rules). In this sense, eController was a

Autonomous Learning Systems: From Data Streams to Knowledge in Real-time, First Edition. Plamen Angelov.

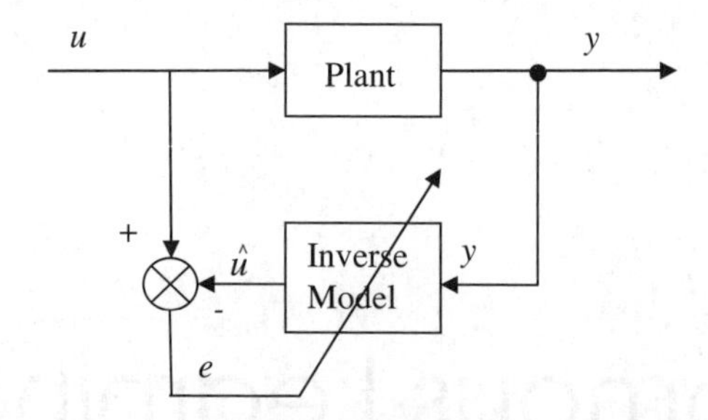

Figure 9.1 A schematic representation of the indirect adaptive control scheme according to Psaltis, Sideris and Yamamura (1988)

pioneering result – the first publication on controllers that adapt their structure (not just tune parameters) online, during the process control and can start 'from scratch' or be initialised. This type of controllers was later further developed by Cara *et al.* (2010) by adding new membership functions (MF) in places where the mean squared error is high and applied to laboratory and industrial processes and mobile robotics (Sadeghi-Tehran *et al.*, 2010). Another significant step forward was made with the AnYa-type fuzzy rules that made possible (antecedent)-parameter-free controllers to be autonomously designed (Sadeghi-Tehran *et al.*, 2012) and (Angelov, Skrjanc and Blazic, 2012).

In the remainder of this chapter the principles of indirect adaptive control scheme will be described first. This will be followed by the description of the inverse plant dynamics model and, finally, the autonomous controller structure.

9.1 Indirect Adaptive Control Scheme

Indirect adaptive control (Andersen, Teng and Tsoi, 1994) is a scheme that aims to substitute the inverse plant dynamic by a model (usually a neural network or a fuzzy rule-based system), Figure 9.1.

If the dynamics of the plant (usually expressed by a set of differential equations) is completely known and available, the control problem reduces to a selection of the controller structure and tuning the parameters. However, in reality, the plant dynamics is often subject to evolution and changes due to wear, change of operating regimes, variations in the quality of raw materials, contaminations, and so on. This leads to models of the plants that become quickly imprecise and noisy.

An alternative is offered by the robust control approach that, essentially, studies the bounds on possible changes to the model of the plant and controller. There are studies for controller adaptation including its structure when such bounds are known in advance (Gao and Er, 2003). In the more general case, when such bounds are not known the structure of the controller remains to be determined.

For example, if in a discretised form, the plant dynamics is denoted by the following function

$$y_{k+1} = f(x_k, u_k); \quad k = 0, 1, \ldots \tag{9.1}$$

where

$x_k = (y_k, y_{k-1}, \ldots, y_{k-p};\ u_{k-1},\ u_{k-2}, \ldots, u_{k-q}) \in Z$ is the state vector;
Z is the operating regime of the plant;
$u_k \in R$ is the control signal exerted by the plant;
f is unknown (generally, nonlinear) function;
p and q are constants that determine the order of the plant.

The controller produces the control signal,

$$u_k = g\left(\zeta_k, a_k\right)$$

where

$\zeta_k = \left(y_k^{\text{ref}}, y_k, y_{k-1}, \ldots,\ y_{k-p}; u_{k-1}, u_{k-2}, \ldots, u_{k-q}\right)$ denotes the extended states vector and differs from the state vector, x by appending it with the reference signal,
y_k^{ref} at the front;
$g(.)$ is the (nonlinear) function of the controller;
a_k is the vector of parameters of the controller.

The aim of the control is to force the plant output to follow the reference signal, y_k^{ref} as close as possible.

The algorithm of the indirect adaptive control (Andersen, Teng and Tsoi, 1994) can be summarised as:

1. start with applying any (possibly, generated by a simple proportional or PI type) control;
2. action, u_1 for one sample, $k = 1$;
3. measure the actual output of the plant **at the next time step**, y_{k+1} after applying u_k;
4. model the inverse plant dynamics by the mapping $u_k = g\left(\zeta_k, a_k\right)$;
5. apply the control signal, u_{k+1} derived using the mapping from the previous step using y_k^{ref} in the state vector;
6. go to step 2 unless a stop criteria is reached.

Using this indirect adaptive control algorithm in order to achieve a certain reference value of the output, y_k^{ref} it is necessary to apply such a control signal u_k which would have caused, ideally, at the next time step an output, $y_{k+1} = y_k^{\text{ref}}$.

9.2 Evolving Inverse Plant Model from Online Streaming Data

The schematic representation of the proposed autonomous controller *AutoControl* can be given in Figure 9.2 below where a simplified state vector that is composed of the error and error difference components, $\xi_k = (e_k, \Delta e_k)$.

One way to train the controller is to use offline learning and apply NN or a FRB because both are proven universal approximators. However, the offline training requires

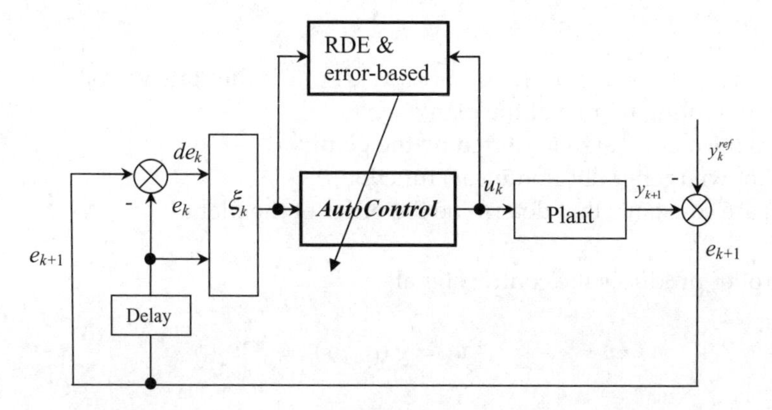

Figure 9.2 A schematic representation of *AutoControl*

availability of a large number of representative training samples and is against the very idea of adaptive control, because it ignores the possibility of the plant dynamics changing.

Equivalently to the use of the plant output directly, one can use the error, e_k (and change of error) formed between the plant output, y_k and the reference value, y_k^{ref} as shown in Figure 9.2.

An effective alternative for online autonomous self-learning is presented by applying ALS of ETS type (Angelov, 2004a; Cara *et al.*, 2010) or AnYa type (Sadeghi-Tehran *et al.*, 2012) with zero order consequents:

$$\begin{aligned} &Rule_i : IF\,(\xi_k \sim \xi_i^*) \\ &THEN\,(u_{ik} = U_{ik}) \end{aligned} \tag{9.2}$$

where

U_{ik} denotes the singleton outputs of the consequent part;

$\xi_k = \begin{bmatrix} e_k \\ de_k \end{bmatrix}$ denotes the state vector composed of the error and error difference components.

The overall output is determined (as with other FRB models of so-called Takagi–Sugeno–Kang type) using centre of gravity defuzzification:

$$u_k = \sum_{i=1}^{R} \lambda_{ik} U_{ik} \tag{9.3}$$

where λ_i denotes the firing strength of the ith local submodel.

In this case, the structure of the controller (fuzzy rules, related focal points) is determined online and autonomously during the process of control itself. In such a scenario, one can either start 'from scratch' and learn 'on the fly' or use some *a priori* simple controller (such as P, PI, FLC) for the initial short period to avoid safety or other possible problems that may be related to the first few (usually, half a dozen or so) samples when the inverse plant dynamics model and, respectively, the controller structure is not yet established. The role of the initial control algorithm (for the first few time steps) is just to generate several starting training samples and is replaced after these initial steps. Therefore, it is not critical which algorithm exactly will be used. Such an approach represents a true *'learning trough experience'*. This flexible and intelligent autonomous controller is called *AutoControl*.

9.3 Evolving Fuzzy Controller Structure from Online Streaming Data

An overall flow chart diagram that represents *AutoControl* is provided in Appendix B8. Its principle of work combines the concepts of indirect adaptive control (described in Section 9.1), of AnYa-type nonparametric (in terms of the antecedents) structure (described in Chapter 5) and the least mean squares approach for tuning the consequents.

AutoControl can start from a very simple configuration or even 'from scratch' and self-develop its structure based on the data obtained online. It has two main phases. In the first phase the local submodels (singletons) are determined and in the second phase the focal points of the data clouds, x_i^*, which are identified.

The modification of the controller structure is driven by the same three principles, A–C described at the beginning of Chapter 7 and used in ALS, in general. The underlying concept is that of the data density determined recursively (using RDE). In practice, different sequences of training data may produce different structures (the approach is order dependent).

AutoControl starts with the first data sample being assumed to be the first focal point, unless an alternative initialisation is provided (see the flow chart of the algorithm presented in Appendix B8). The first rule has a form of expression (9.2) and can form its antecedent (IF) part around the data sample x_1. The consequents of the rules are tuned using the LMS-like algorithm. Aiming to minimise the square error, $E = e^2$ LMS follows the gradient-based optimisation:

$$U_{ik} = U_{ik-1} - \frac{\partial E_k}{\partial u_k} \tag{9.4}$$

where the error, $e_k = y_k^{\text{ref}} - y_{k-1}$.

If we denote the update of the outputs per local submodel by $\Delta U_{ik} = U_{ik} = U_{ik-1}$ one can get the equation for the update of the consequent singletons of *AutoControl* using the LMS algorithm, taking into account the expression of the gradient of the error as follows:

$$\Delta U_{ik} = C\lambda_{i(k-1)}e_k \tag{9.5}$$

where C is a constant.

Additionally, in many practical applications there are limits over the control signal. Let us denote them by $u_{\min}$ and $u_{\max}$. Then, additional actuators limitations can be imposed (Sadeghi-Tehran *et al.*, 2012):

$$\Delta U_{ik} = \begin{cases} 0 & \textit{if}\ (u_{k-1} = u_{\min})\ AND\ (\Delta U_{ik} < 0) \\ 0 & \textit{if}\ (u_{k-1} = u_{\max})\ AND\ (\Delta U_{ik} > 0) \\ \Delta U_{ik} & \text{otherwise} \end{cases} \tag{9.6}$$

It should be stressed that the consequent adaptation is performed online while the controller is operating; therefore, the control action is applied from the very first moment.

9.4 Examples of Using *AutoControl*

Two simple examples are shown here with *AutoControl* primarily with illustrative purposes. The first example is from the area of food (sugar production) industry. The aim is to control the changing reference point of the water level in the tank (Sadeghi-Tehran *et al.*, 2012). The discretised differential equation (of first order) that describes the process can be given by:

$$y_{k+1} = y_k + T\left(\frac{-\sqrt{19.6y_k}}{y_k^2+1} + \frac{u_k}{y_k^2+1}\right) \tag{9.7}$$

where T is the sampling rate set to 0.5 s.

In order to simulate the plant one also needs to take into account the obvious physical constraints that the water level is non-negative. The combined equation is:

$$y_{k+1} = \max\left\{0,\ y_k + T\left(\frac{-\sqrt{19.6y_k}}{y_k^2+1} + \frac{u_k}{y_k^2+1}\right)\right\} \tag{9.8}$$

For this simulation the reference point is assumed to be changing by the following law:

$$y_k^{\text{ref}} = \cos(0.05k) + \sin(0.07k) + 3.7 \tag{9.9}$$

Following the procedure described in Appendix B8 and Sections 9.2 and 9.3 *AutoControl* starts with no model and an empty rule base and develops autonomously a structure that consists of three data clouds with prototypes as shown below:

$$\xi^* = \begin{bmatrix} e & de \end{bmatrix}^T = \begin{bmatrix} 4.735 & 0 \\ 1.486 & -0.367 \\ -0.438 & 0.338 \\ 0.632 & 0.434 \end{bmatrix}^T$$

Based on these prototypes, four linguistic rules of the form (9.2) were formulated and the parameters tuned using the LMS-like approach as described above that at the end of the process get values:

$$U_{3000} = \begin{bmatrix} 12.19 \\ 36.09 \\ -44.59 \\ 61.83 \end{bmatrix}$$

It should be stressed that the values of U change all the time.

The evolution of the controller structure is sketched in Figure 9.3 and the performance is illustrated in Figure 9.4.

The position of the focal points and all other data points of the four *clouds* are represented in Figure 9.5.

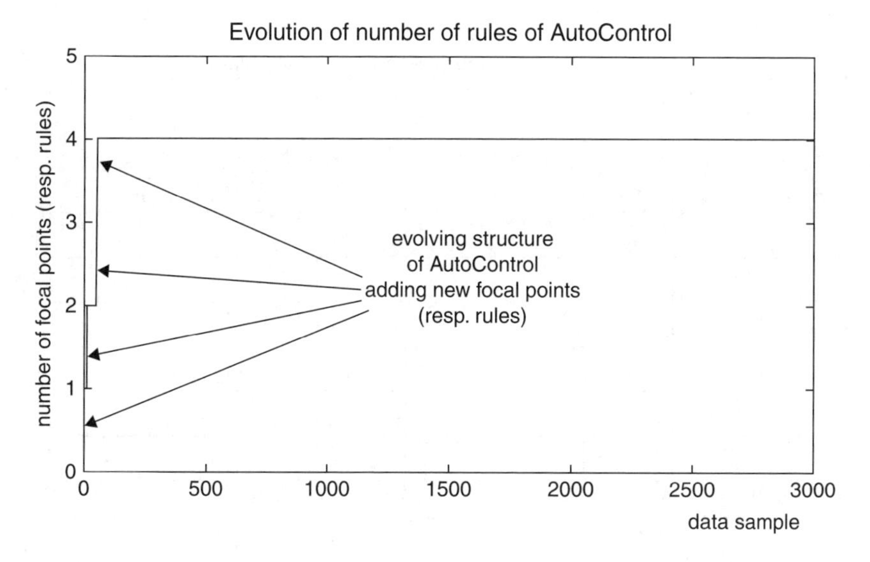

Figure 9.3 Autonomous evolution of the structure of *AutoControl* (adding new focal points and new rules for the sugar tank water level control illustration problem)

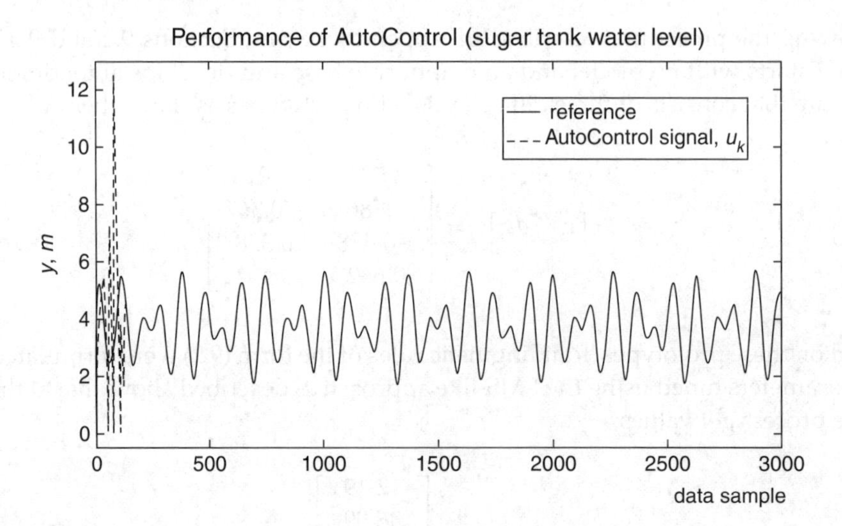

Figure 9.4 Performance of *AutoControl* for the sugar tank water level control illustration problem

It is clear that the controller after a relatively short period of setting its structure completely autonomously 'from scratch' forms its rules with no parameters in the antecedent (IF) part and learning the consequents autonomously collecting feedback from the performance in a closed loop. Yet the performance after the initial period is quite satisfactory (this can be appreciated from Figure 9.4 after data sample 140 onwards till the end, data sample 3000).

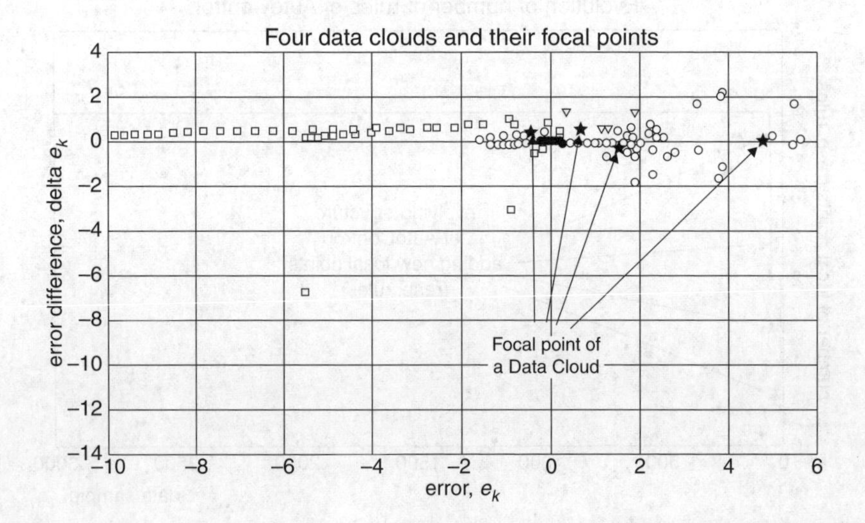

Figure 9.5 Position of the focal points and the data clouds at the end of the control process (after data sample 3000)

The second example represents a control problem in a water bath (Sadeghi-Tehran *et al.*, 2012). The discretised first-order differential equation that describes the process is given by:

$$y_{k+1} = a(T)y_k + \frac{b(T)u_k}{1 + e^{0.5y_k - \gamma}} + (1 - a(T))Y_0 \tag{9.10}$$

where

$$a(T) = e^{-\alpha T};\ b(T) = \frac{\beta(1 - e^{-\alpha T})}{\alpha}$$

$$\alpha = 10^{-4};\ \beta = 0.0087;\ \gamma = 40$$

$$Y_0 = 25\ ^{\circ}\mathrm{C};\ T = 25\ \mathrm{s}$$

The reference signal is a random step-wise function. *AutoControl* again started with no *prior* model and an empty rule base and developed autonomously a structure that consisted of three data clouds with prototypes as shown below:

$$\xi^* = \begin{bmatrix} e & de \end{bmatrix}^T = \begin{bmatrix} 0.405 & 0 \\ 1.086 & 0.314 \\ 0.007 & -0.002 \end{bmatrix}^T$$

Based on these prototypes three linguistic rules are formulated at the end of the process:

$$\begin{aligned} &Rule_1 : IF\left(\begin{bmatrix} e_k \\ de_k \end{bmatrix} \sim \begin{bmatrix} 0.405 \\ 0 \end{bmatrix}\right) THEN\,(U_{1k} = 3.673) \\ &Rule_2 : IF\left(\begin{bmatrix} e_k \\ de_k \end{bmatrix} \sim \begin{bmatrix} 1.086 \\ 0.314 \end{bmatrix}\right) THEN\,(U_{2k} = 0.038) \\ &Rule_3 : IF\left(\begin{bmatrix} e_k \\ de_k \end{bmatrix} \sim \begin{bmatrix} 0.007 \\ -0.002 \end{bmatrix}\right) THEN\,(U_{3k} = -1.318) \end{aligned} \tag{9.11}$$

The evolution of the controller structure is sketched in Figure 9.6 and the performance is illustrated in Figure 9.7.

The position of the focal points and all other data points of the three clouds are represented in Figure 9.8.

It is clear that the controller, after a relatively short period of setting its structure, completely autonomously 'from scratch' forms its three rules with no parameters in the antecedent (IF) part and learning the consequents autonomously collecting feedback from the performance in a closed loop. Yet the performance after the initial period is quite satisfactory, which can be appreciated from Figure 9.4 after data sample 90 and then after changing the reference point around data sample 670 onwards till the end (data sample 5000).

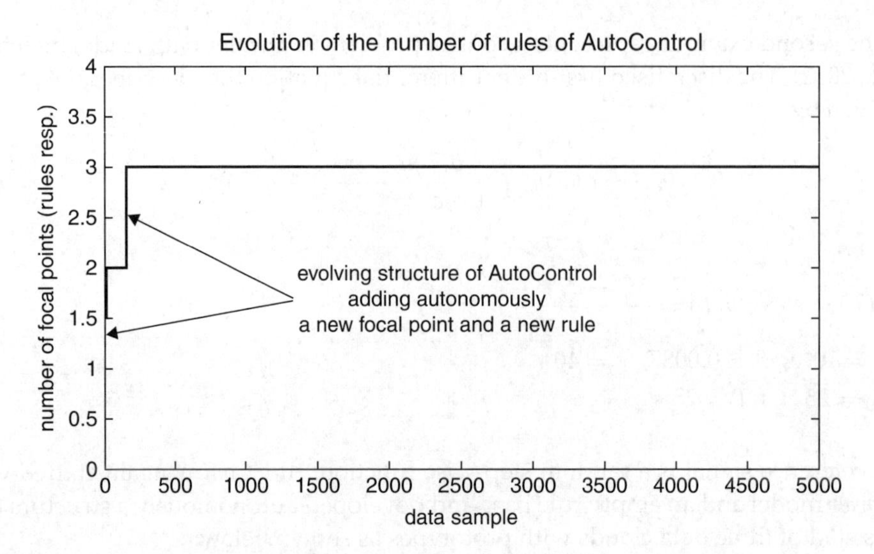

Figure 9.6 Autonomous evolution of the structure of *AutoControl* (adding new focal points and new rules for the temperature control illustration problem)

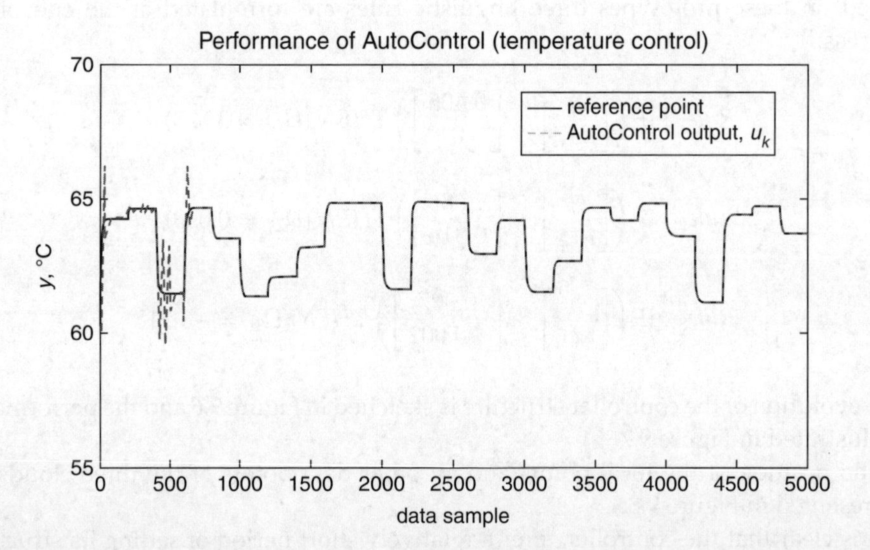

Figure 9.7 Performance of *AutoControl* for the temperature control illustration problem

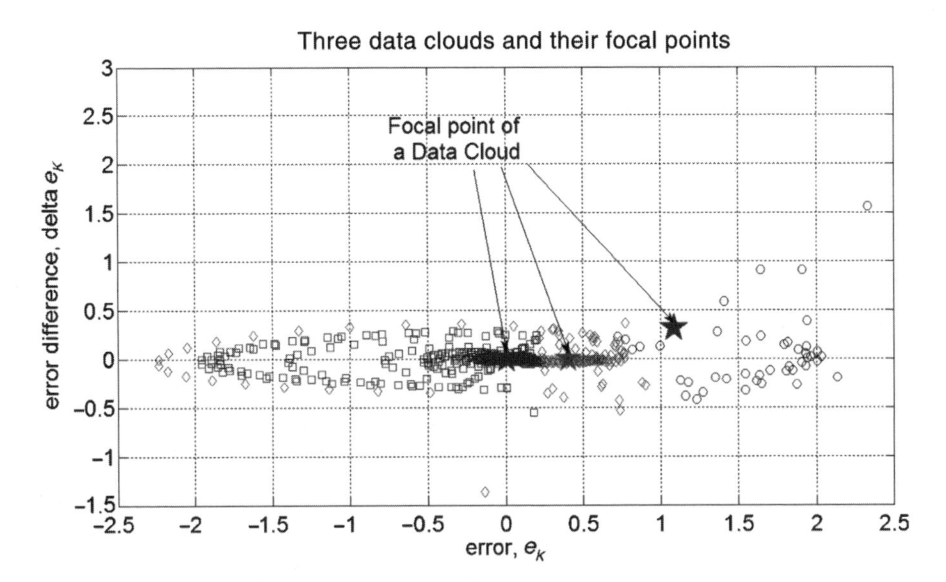

Figure 9.8 Position of the focal points and the data clouds at the end of the control process (after data sample 5000)

9.5 Conclusions

The autonomous learning controller, *AutoControl* described in this chapter has the following distinguishing features:

- It does not require the model of the plant to be known.
- It does not require previous knowledge about the control policy to be known.
- The controller structure is self-developed (possibly starting 'from scratch') based on the density and error information from the history of control process collected during its operation and used recursively (without memorising, but using it fully).
- *AutoControl* has high adaptive ability and corrects later the initial rules when necessary based on experimental data collected during the process run.

10

Collaborative Autonomous Learning Systems

In many applications there may be more than one systems that can act autonomously. There are numerous examples of such situations. For example, students and teachers in a class room, drivers and pedestrians on the roads, and so on. These are natural systems. Similarly, one can imagine artificial autonomous systems (robots, agents). For example, a team of uninhabited vehicles (aerial, ground, water or underwater), a sensor network in which each node may be an *intelligent* sensor, enabled with processing and communication capability, ensemble of classifiers that may just be a software or including some hardware realisation (e.g. image-based), and so on. In such scenarios there may be, generally, the following two modes in which these autonomous entities work, as follows:

a. completely independent or competitive;
b. cooperative or collaborative, which itself can be:
    i. centralised;
    ii. decentralised.

In what follows the collaborative scenario will be briefly described as applied to ALS in the form of clustering, classifiers, controllers, predictors, estimators, filters or intelligent sensors.

10.1 Distributed Intelligence Scenarios

In a collaborative scenario, each ALS acts on its own pursuing its own objectives and goals, but they can collaborate to achieve a common, shared goal. For example, in the so-called self-localisation and mapping (SLAM) problem (Choset and Nagatani, 2001)

Autonomous Learning Systems: From Data Streams to Knowledge in Real-time, First Edition. Plamen Angelov.

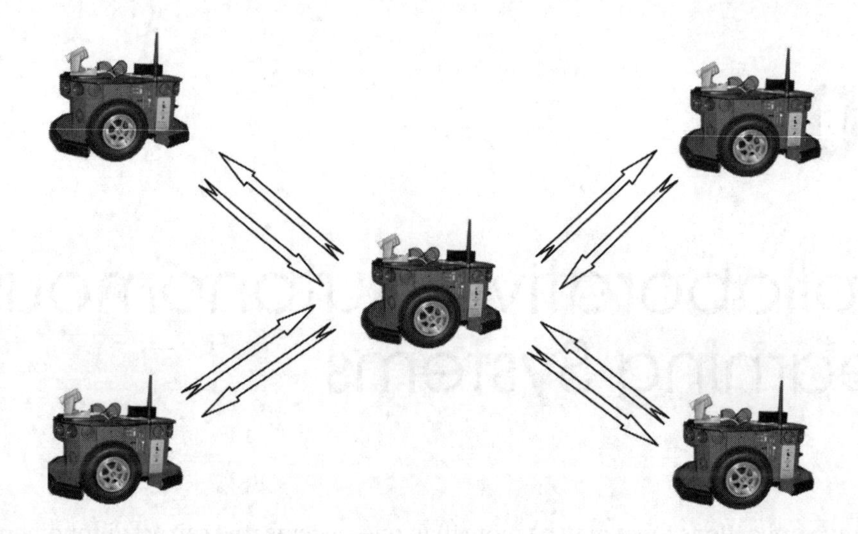

Figure 10.1 A schematic representation of a system of M centralised collaborating ALS

each mobile robot from a team can act on its own to localise and start mapping the environment (an example of application of ALS for such a task will be described in Chapter 12), but the mobile robots may also, in addition to this individual behaviour, exchange the aggregated information they have discovered (e.g. landmarks, their location, possibly imagery, relative position and some additional data that are essential for continuing to explore the environment), Figure 10.1. In a collaborative scenario, there are $M > 1$ ALS (which may be implemented on a dedicated hardware, such as mobile or static sensor platform, mobile phone, laptop computer, mote – mobile node of a sensor network, for example SunSpot by Sun Corporation (Healy, Newe and Lewis, 2008), and so on or simply as separate software agents, where M is an integer number.

There can, broadly, be two different modes of operation:

Mode 1 Centralised collaborative ALSs
In this mode, there is a *Leading* ALS and members of the team. The difference is that all ALS send information to the *Leading* ALS from time to time, periodically or when requested to do so. It can also send back requests or information.

The advantages of the centralised mode are that less information needs to be exchanged reducing the bandwidth requirements and probability of interception in unmanned aerial/ground-based/undersea vehicles (UxV) application for reconnaissance and surveillance, for example.

However, a serious disadvantage of this mode is the vulnerability – if the *Leading* ALS is damaged or disappears, the whole team will have reduced access to previously collected information (which was collected only by the *Leader* and distributed to other members of the team). Since each member of the team is an ALS as opposed to a

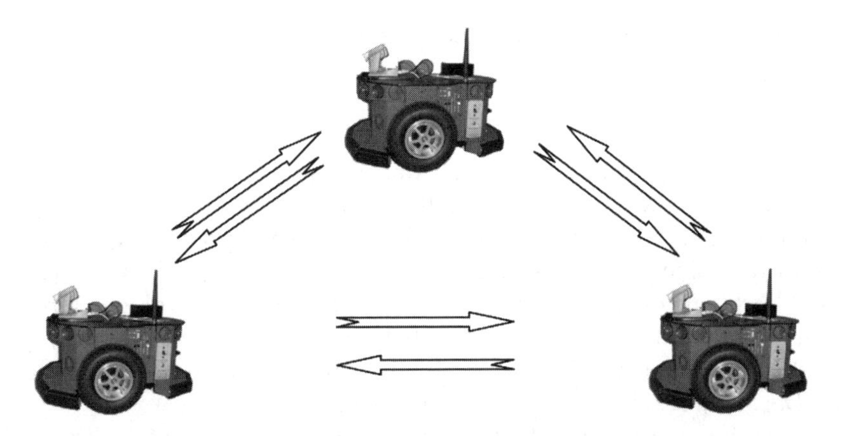

Figure 10.2 A schematic representation of a system of *M* decentralised collaborating ALS

fixed structure system this has a lesser effect and, therefore, one can recommend either applying a decentralised mode from the very beginning or switching to a decentralised mode in case the *Leading* ALS is destroyed or disappears.

Mode 2 Decentralised collaborative ALSs
In this mode there is no *Leading* ALS and all ALSs are equal in terms of exchange of information. This mode is depicted in Figure 10.2. In what follows only the decentralised mode will be considered.

10.2 Autonomous Collaborative Learning

Autonomous systems can collaborate while performing their mission in terms of any or a combination of the following tasks (which have been described and discussed earlier in the book; see Chapters 7–9):

a. density estimation and clustering of the incoming sensory data;
b. prediction, filtering, estimation, self-calibrating inferential sensors;
c. classification;
d. control.

In addition, collaborative decision making and situation awareness (Endsley, 1996) is of great interest to UxVs and can benefit from some of the above, but will not be specifically considered in this book.

The following interesting ***hypothesis*** can be formulated:

Let us assume that there are $M > 1$ collaborating ALS that take streaming data as input and perform either of the tasks a–d as described above. Let us denote the multivariate, possibly nonstationary, data stream by DS. Let us assume it is composed of chunks (parts) as follows:

$$DS = [DS_1, DS_2, \ldots, DS_K];\ K > 1$$

Let us assume that one ALS processes the data stream DS and let us denote the result by *Result*. Let us now assume that another ALS gets (receives, possibly transmitted/communicated, possibly, wirelessly or simply reads previously stored) certain amount of recursively calculated interim set of variables, V_j (where $j = [1, K]$) and after that starts processing the data stream from the $DS_{j+1}{}^{\text{th}}$ chunk onwards. For example, if $j = 1$ that means that the ALS gets V_1 and data stream $[DS_2, DS_3, \ldots, DS_K]$ and the result is denoted as $Result_1$; if $j = 3$ that means that the ALS gets V_3 and data stream $[DS_4, DS_5, \ldots, DS_K]$ and the result is denoted as $Result_3$.

The interesting question is – *if we compare the results (Result, $Result_1$, $Result_3$) how do they relate to each other?*

Angelov (2006) has proven that for any of the tasks *a–d* they will be *exactly* the same! Because the ALS methods described in this book are recursive and use ***all*** historical data ***without memorising them*** this leads to the very important and useful conclusion that an ALS (e.g. mobile robots or agents) can exchange only V_j and in this way guarantee that the result (in terms of classification of a target or clustering, density estimation or prediction and so on as listed above) will be ***exactly*** the same made by another ALS that never has seen a part of the data stream $[DS_1; \ldots, DS_j]$. This leads to huge savings in bandwidth because there is no need to transmit the raw data ($[DS_1; \ldots, DS_j]$) yet the same conclusion can be made as if the data was available to all ALS.

This method is more efficient than simply a compression that is, usually, not exact. The rate of reduction of the transmitted information is given by the ratio $\xi = \frac{\dim(V_j)}{\dim([DS_1; \ldots; DS_j])}$ where dim(.) denotes the dimensionality of the vector. A realistic example would be $\dim(Vj) = 50$; $\dim([DS_j, \ldots, DS_K] = 10\,000$. For such an example, $\xi = 200$, but much bigger savings are also possible in a real-time scenario with imagery data, for example. Moreover, one can get ***exactly*** the same result as if 200 or more times more data has been processed and transmitted!

10.3 Collaborative Autonomous Clustering, *AutoCluster* by a Team of ALSs

For example, a team of collaborative ALSs for clustering can be demonstrated by a collaborative SLAM task:

The *focal points* are determined online by an ALS (Figure 10.3) as described in Section 3.2 using RDE. The aggregated information as described below can be transmitted to the other ALSs. Therefore, the vector of the data that has to be transmitted between the ALS includes:

- the focal points, x^*_{ij} where j is the time instant when the data is transmitted;
- the statistical variables needed to continue the recursive calculation of the RDE: D^*_{ij}, μ_j, $\sum_j$.

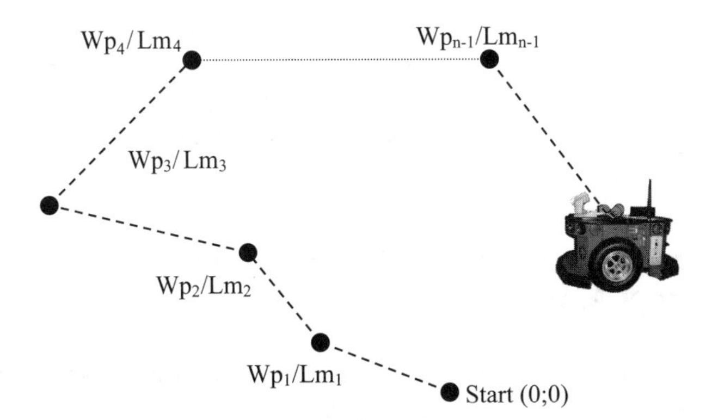

Figure 10.3 A schematic representation of an ALS performing clustering. Wp – way point; Lm – landmark. This may be a part of a collaborative SLAM task if other ALSs are involved and aggregated information is passed to them

Therefore, for collaborative clustering, $V_j = [x_{ij}^*;\ D_{ij}^{i*};\ \mu_j;\ \Sigma_j]$. Their dimension is $4Rn$:

R coordinates of the focal points with dimension n,
R densities (of focal points),
R mean values and
R covariance values – all with the same dimension.

For example, if $R = 4$ and $n = 3$ (e.g. R,G,B in a colour-coding scheme) that makes 48 (real) variables. They may represent thousands or millions of raw data (3D in this case) of a data stream.

10.4 Collaborative Autonomous Predictors, Estimators, Filters and *AutoSense* by a Team of ALSs

In this scenario, the *focal points* are determined online as described in Chapters 5 and 7 that includes, in general, two subtasks:

i. structure evolution: and
ii. parameter learning.

Therefore, the vector of the data that has to be transmitted between the ALS includes:

- the focal points, x_{ij}^* where j is the time instant when the data is transmitted with dimension Rn;
- the statistical variables needed to continue the recursive calculation of the RDE: D_{ij}^*, μ_j, $\sum_j$ each with dimension Rn;
- local submodels parameters: A_j with dimension $R(n+1)$.

Therefore, for collaborative prediction, filtering, estimation or self-calibrating sensors, $V_j = [x^i_{ij}; D^*_{ij}; \mu_j; \Sigma_j; A_j]$. The dimension of the vector that is to be transmitted is $5Rn + R$. For example, if $R = 4$ and $n = 3$ (e.g. R,G,B in a colour-coding scheme) that makes 64 (real) variables. They again may represent a data stream with thousands or millions of raw data (3D in this case).

10.5 Collaborative Autonomous Classifiers *AutoClassify* by a Team of ALSs

For example, a team of collaborative ALSs for the autonomous classification task can be illustrated by the following example (Figure 10.4):

In this scenario, the *focal points* are determined online as described in Chapters 5 and 8. Therefore, the vector of the data that has to be transmitted between the ALS includes:

- the focal points, x^*_{ij}, where j is the time instant when the data is transmitted with dimension Rn;
- the statistical variables needed to continue the recursive calculation of the RDE: D^*_{ij}, μ_j, $\sum_j$ each with dimension Rn;
- if *AutoClassify1* type of ALS is used then, in addition, the local submodels parameters: A_j with dimension $R(n + 1)$ are also included.

Therefore, the amount of data transmitted in collaborative autonomous classification is the same as in clustering if *AutoClassify0* is used and the same as predictive/

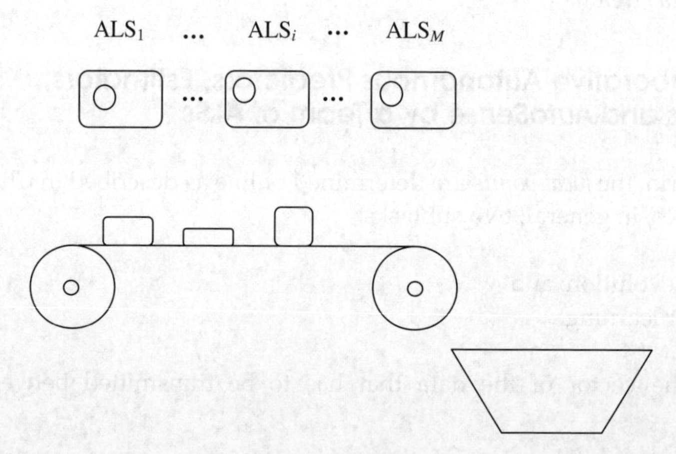

Figure 10.4 A schematic representation of collaborative ALSs performing autonomous classification. Each camera may classify the image data into separate objects (biscuits, CDs, etc.) into 'good' or 'bad' automatically. ALS_1 may learn new rules online and may pass this information to ALS_2

estimation/filtering or self-calibrating sensors if *AutoClassify1* is used – see previous two subsections for details.

10.6 Superposition of Local Submodels

In a collaborative scenario when an ALS receives the vector V_j with the recursively calculated variables it continues the autonomous learning process using the information for initialisation. If the receiving ALS has already generated a model then a superposition of the two models will take place adding the transmitted focal points (and pdf if use probabilistic model) to the existing focal points (and, respectively, membership functions or pdfs if relevant). This may raise problems of interpretability and may require model simplification as described in Section 5.6. Note that this is only relevant in the case of Mamdani or Takagi–Sugeno FRB. If an AnYa-type systems is used there is no need for membership functions and, respectively, their parameters.

10.7 Conclusions

In this chapter the powerful and interesting idea of the team of ALS that can collaborate is briefly described. It was first described by Angelov (2006) in a patent application and tested with the ETS algorithm (Angelov and Filev, 2004). It has huge unexploited potential in areas such as mobile robotics, wireless sensor networks and unmanned vehicles (UxVs) to name a few. It allows reduction of the complexity of problems by reducing it to simpler subproblems, dramatically reducing the amount of information required to be transmitted and, thus, communication bandwidth). It also allows increased survivability in multi-UxV tasks by performing collaborative task execution without loss of information by an UxV that may be affected.

PART III

Applications of ALS

11

Autonomous Learning Sensors for Chemical and Petrochemical Industries

One of the most interesting applications of ALS for prediction and estimation is for self-calibrating (*intelligent*/*smart*/*soft*/inferential) sensors. They are applicable in various industries, but most widely in chemical and petrochemical branches. In this chapter there will be some illustrative examples of the research work the author did collaboratively with Dr. Jose Macias Henrandez from CEPSA oil refinery, Santa Cruz de Tenerife, Spain and Dr. Arthur Karl Kordon, The Dow Chemical, Freeport, Texas, USA.

11.1 Case Study 1: Quality of the Products in an Oil Refinery

11.1.1 Introduction

In oil refineries the crude oil is separated in different (quality) petroleum cuts (being refined) in distillation towers by lateral extraction (Macias and Feliu, 2000; Macias, Angelov and Zhou, 2006), see Figure 11.1.

These cuts include:

- gasoline;
- gas oil;
- naphtha;
- kerosene, and
- other commercial products.

Autonomous Learning Systems: From Data Streams to Knowledge in Real-time, First Edition. Plamen Angelov.

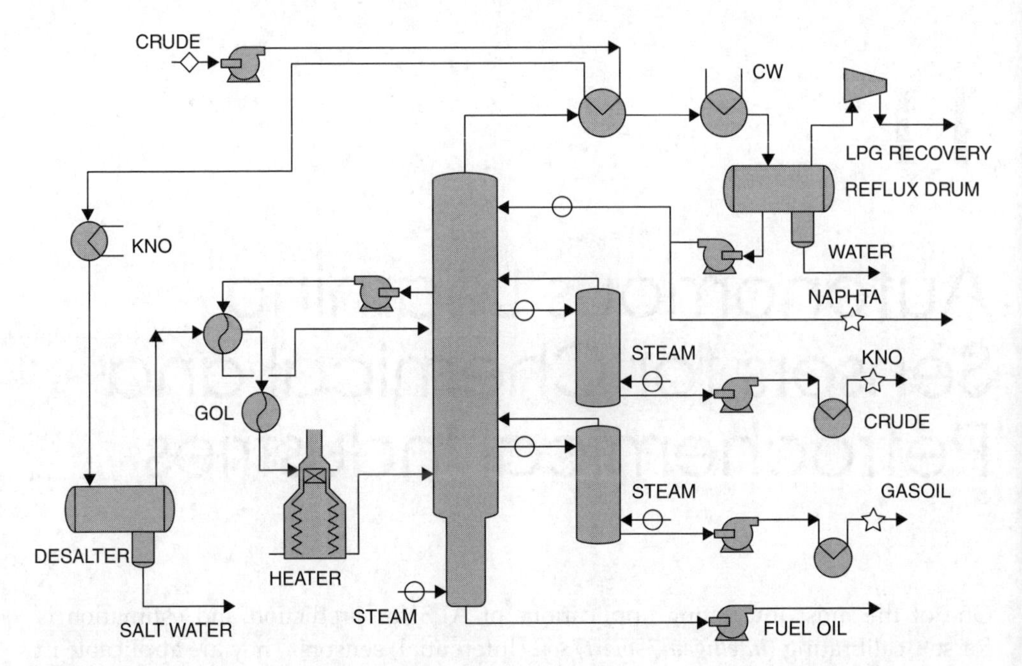

Figure 11.1 A schematic representation of a distillation unit (adapted from Macias-Hernandez and Angelov from Angelov, Filev and Kasabov © John Wiley & Sons, Ltd., 2010). The points at which the data are measured periodically (with high frequency) using conventional ('hard' sensors) are indicated with circles. Points at which the product is taken and its quality is to be predicted are indicated with stars and the point at which the crude enters the distillation unit is indicated by a diamond

Specification requirements for these technological processes are very strict, because the yield obtained in these cuts contributes significantly to the overall refinery profit. The amount of product that can be extracted in a side stream is regulated and limited by a voluntary standard called ASTM (ASTM, 2011). For example, the so-called '95% ASTM' distillation curve analysis is the relation between the vapour leaving the pot and the distillated product and temperature.

The commercial aim is, obviously, to obtain maximum amount of product with the highest quality within the specification. Extracting more of the product from distillation towers is linked to higher end points and heavier cuts. Maintaining the quality and specification is, therefore, a very important goal of the oil refining process.

11.1.2 The Current State-of-the-Art

Currently, expensive and time-consuming laboratory analysis is performed to the side stream products in order to monitor the quality. In some refineries they install continuous analysers, but this is an even more costly solution (Hernandez and

Angelov, 2010). It is often difficult to maintain a stable correlation between the results from the laboratory analysis and the continuous analysers.

They started using so-called *inferential* sensors to estimate the product quality from available plant measurements, such as temperature, flow rates, pressures as far back as 1960s. Later, they started to also use online analysers and statistical models such as PCA and PLS (partial least squares), which are currently widely applied. Later, in the 1990s, came NN-based inferential/*soft* sensors.

All these methods suffer from errors in the measurements as well as from the sensor/model structure inadequacy. In reality, the plant (distillation tower) has a dynamic, nonstationary behaviour, the crude oil characteristics are variable, contaminations are often present, but the model is fixed. In addition, effects such as '*drift*' and '*shift*' of the data stream also cannot be taken into account by fixed model structures (Angelov and Kordon, 2010).

11.1.3 Problem Description

The data used for this study is from a crude distillation unit with a design capacity of 80 000 barrels per day of crude from the Arabian Peninsula. The distillation tower has valve trays and two cold pumps around, kerosene oil (KNO), gas oil (GOL) and a hot pump around GOP (washout). It has five side streams from top to bottom listed below (Hernandez and Angelov, 2010) and a vapour-side stripper for each side stream and a bottom vapour injection:

- heavy naphtha (HN);
- kerosene oil (KNO);
- light gas oil (LGO);
- medium gas oil (MGO);
- heavy gas oil (HGO);
- atmospheric residue (RES).

11.1.4 The Dataset

The data (courtesy of Dr. Jose Macias Hernandez, CEPSA Oil Refinery, Santa Cruz de Tenerife, Spain) include the analysis from the laboratory (usually performed once a day, seven days a week) and the 'hard' sensors data from the tower operation. The readings from hard sensors are taken much more often, but in order to be compatible with the limitation that the laboratory results impose they are averaged daily. The data include the period of the whole 2006 starting on 1 January and ending 31 December. During this period there were emergencies, shutdowns, process and instrument malfunction and even laboratory sample errors. No filtering was applied to the dataset (it is real and raw).

The laboratory analysis data is, in fact, the training (target) output data, y including the oil fractions listed above (HN, KNO, LGO, MGO, HGO). The instrument readings of temperatures, flow rates and pressures of the main tower are, in fact, the input data vector, x.

Let us denote by y the output variable of interest, for example the temperature of the heavy naphtha when it evaporates 95% liquid volume (T^{hn}) or the temperature of the gasoil (T^{gol}). The aim is to model it as a function of the measurable variables, x, such as the pressure at the tower (p), the amount of the product taking off (P), the density of the crude (ρ), the temperature of the column overhead (T^{co}), the temperature of the naphtha extraction (T^{ne}).

The most important variables in the monitoring and analysis of a distillation process are the extraction temperatures of different fractions (products). The physical variables, which are easy to measure (such as density of the crude, amount of product taken off, pressure in the tower, etc.) are highly correlated.

The error (which represents a performance indicator for *AutoSense*) is based on the standard deviation of the absolute errors in regards to the average error for the period of recalibration:

$$e^2 = \left(\frac{1}{N_{val} - 1}\right) \sum_{i=1}^{N_{val}-1} (\varepsilon_i - \overline{\varepsilon})^2 \tag{11.1}$$

where

e is the error used by *AutoSense*;
N_{val} is the number of samples during the calibration period;
ε denotes the absolute error;
$\overline{\varepsilon}$ denotes the mean/average error.

11.1.5 AutoSense *for Kerosene Quality Prediction*

For example, one of the problems of interest is to predict the temperature of the kerosene, T^k (°C) when it evaporates 95% liquid volume according to the standard ASTM D86-04b (ASTM, 2011). It depends mostly on the following factors (physical variables that are easily measurable by traditional/'hard' sensors with high frequency):

- pressure of the tower, p (kg/cm^{2}g);
- amount of the product taken off, P (%);
- density of the crude, ρ (g/l);
- temperature of the column overhead, T^{co} (°C);
- steam introduced in gasoil, SGK (kg/h);
- temperature of the kerosene, T^{ke}, (°C).
- temperature of the naphtha extraction, T^{ne}, (°C).

The results applying *AutoSense* are depicted in Figure 11.3. *AutoSense* recalibrates autonomously, reducing lifecycle costs related to maintenance, daily laboratory tests, human involvement (including lab technician and chemical process engineer).

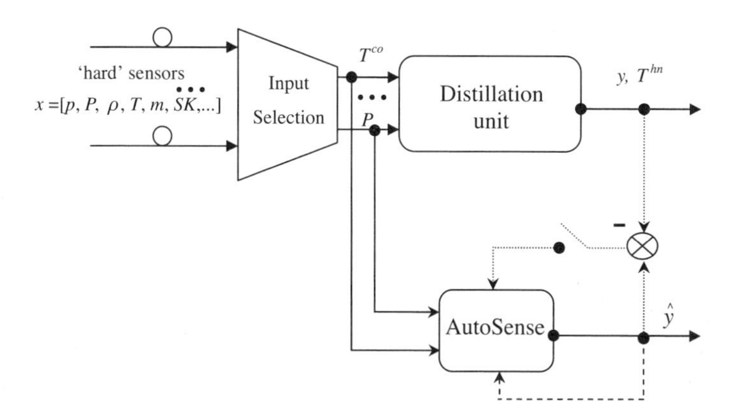

Figure 11.2 A schematic representation of *AutoSense* for the oil refinery. Notations concern an example of predicting the temperature of the heavy naphta, T^{hn} when it evaporates 95% liquid volume according to the standard ASTM D86-04b

AutoSense can theoretically start 'from scratch', but it is more practical to initialise it with a calibrated *soft* sensor. This can be done, for example, by using *AutoSense* to predict hourly or more often if needed, but on a daily basis recalibrate (automatically) by providing the 'true' value of T^k. By 'true' value here we mean the result of the laboratory test (which itself relies on using graphs and first-principles models). This initial training period (during which *AutoSense* still predicts, but requires on a daily basis the 'true' values to be provided) can last (as was the case in our experiment) for some 90 days (3 months). After this initial period *AutoSense* does not require any human intervention and works, for example, for the next 6 months (180 days) in a completely unsupervised manner. After this 180-day period a relatively short retraining/recalibration period of 60 days again requires 'true' values of the variable that is predicted, T^k.

However, it should be stressed that even during this period there is no need for human intervention and the retraining is done recursively, so there is no complete remodelling and iterative processes. Following this retraining period, *AutoSense* predicted for further 4 months (120 days) completely autonomously. In this way, retraining periods are only required from time to time, but there is no need for expert involvement to remodel the sensor. This is indicated in Figure 11.2 by dotted lines.

Internally, *AutoSense* evolves its structure as described in Chapter 7 by adding or removing local submodels that is visualised in Figure 11.4.

From the figure it can be seen that the internal stricture of *AutoSense* is very simple and dynamically evolves. The overall model is nonlinear (even if local linear submodels are used), nonstationary (because it is evolving) and non-Gaussian (even if Gaussian local submodels are used). In fact, *AutoSense* has a dynamically evolving multimodel structure that is not prefixed.

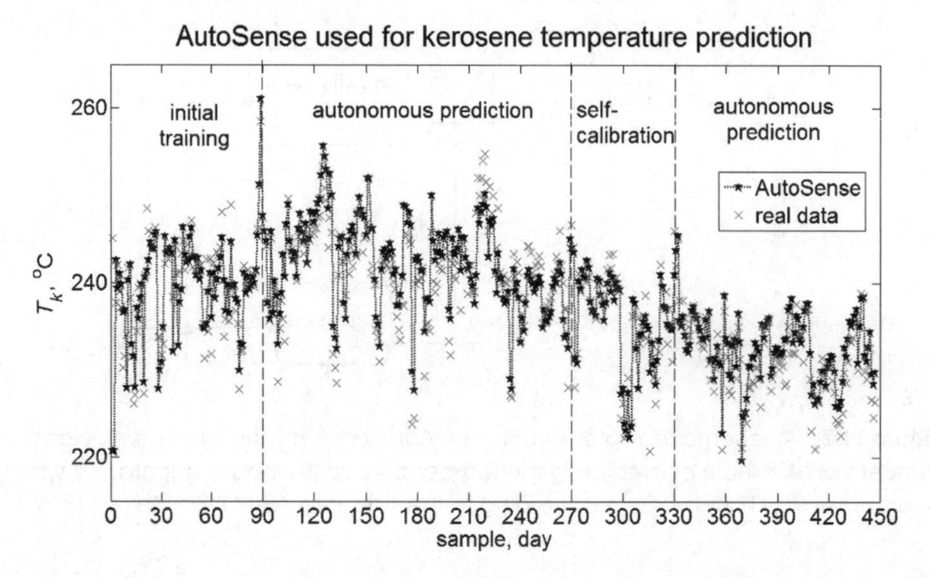

Figure 11.3 Results of prediction of the kerosene temperature when it evaporates 95% (according to ASTM D86-04b); the data from the laboratory analysis are shown with crosses; the data from the predictions are shown with stars linked with a dotted line. The value of the error, *e* is below 2%

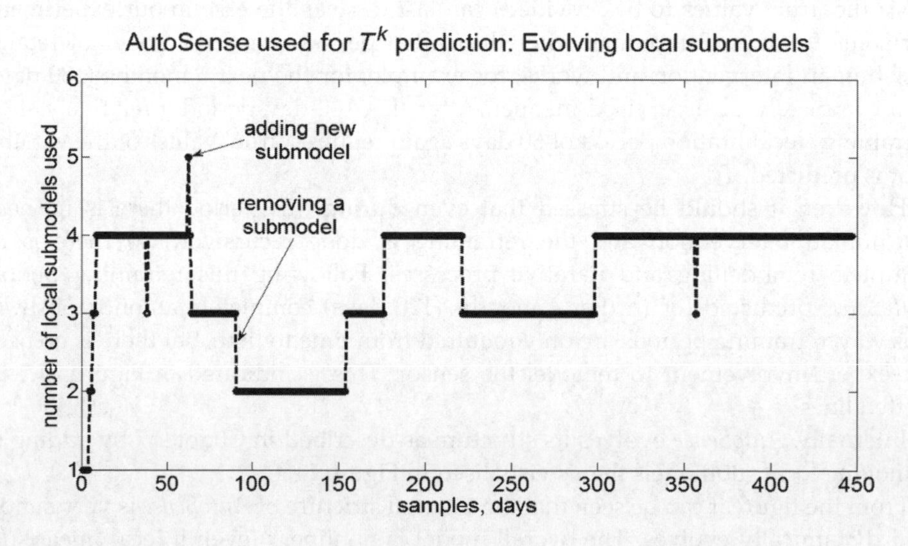

Figure 11.4 Evolution of the local submodels in prediction of temperature of kerosene when it evaporates 95% (according to ASTM D86-04b) problem described above

11.1.6 AutoSense *for Abel Inflammability Test*

A problem of high importance not only as an indirect measurement of physical variables, which is difficult or impossible to measure directly, but also from the point of view of safety is the so-called inflammability index.

This depends mostly on the following factors (physical variables that are easily measurable by traditional/'hard' sensors with high frequency):

- pressure in the tower, p (kg/cm^{2}g);
- amount of the product taken off, P (%);
- density of the crude, ρ (g/l);
- temperature of the column overhead, T^{co} (°C);
- steam introduced in kerosene stripper, SK (kg/h);
- Temperature of the naphtha extraction, T^{ne}, (°C).

The results applying *AutoSense* are depicted in Figure 11.5. A value of the error, e as low as 2.23% was reported by Macias-Hernandez and Angelov (2010). In addition, if we use FRB as a framework, transparent and human-understandable rules can be achieved that have linear consequent parts indicating proportionalities that are valid locally around the focal points.

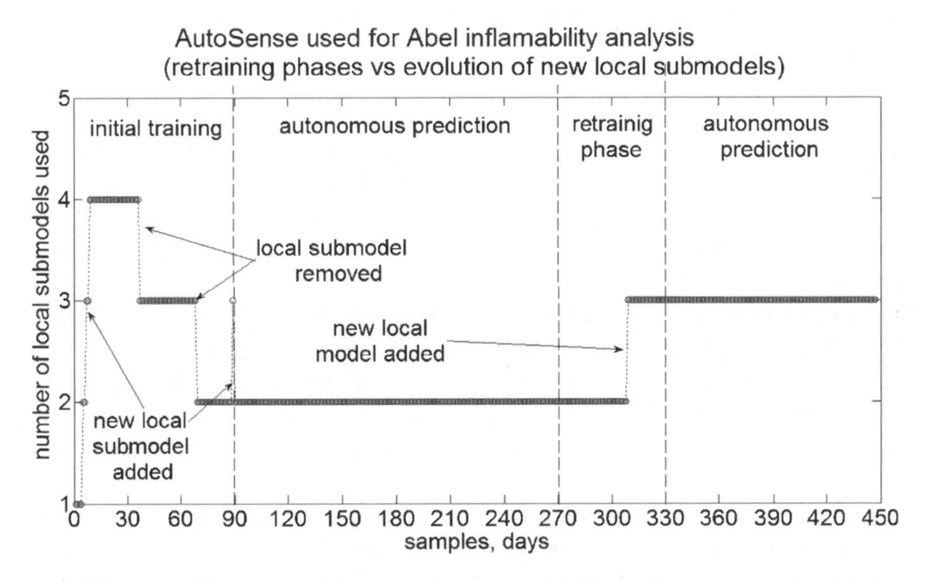

Figure 11.5 Operation of *AutoSense* in terms of retraining phases and evolution of the internal structure (local submodels) for Abel inflammability index prediction in kerosene extraction

11.2 Case Study 2: Polypropylene Manufacturing

11.2.1 Problem Description

Another case study includes polymerisation processes from the chemical industry (courtesy of Dr. Arthur Kordon from The Dow Chemical, TX, USA). *AutoSense* was applied for prediction of the properties of chemical compositions and propylene in a simulated online mode (Angelov and Kordon, 2010). The data were noisy and with incorporated changes of the operating regime of the process. In addition, the number of measured (with *'hard'*/physical sensors) variables is large and input selection is required as a preprocessing or as a part of the overall algorithm. *AutoSense* demonstrated its flexibility in addressing all of these problems that are typical for this real industrial data.

The first subproblem concerns product composition in *the bottom of* the distillation tower with significant amount of noise in the data. The input data include 47 variables (some of which may be correlated and loosely related to the product composition), $x \in R^{47}$. The target output, y is derived from laboratory analysis, which is, however, less noisy than the output for the other three datasets. This dataset also contains a significant operational change around data sample 113.

The second subproblem describes propylene in the top of the distillation tower. In this case, the input data contains 3000 measurements from 22 'hard' sensors, $x \in R^{22}$ taken every 15 minutes using gas chromatography, which cover a very broad range of operating conditions.

AutoSense can be (optionally) initialised but can also self-learn the model from the very first data sample it reads. The prediction can start from the second data sample (this depends on the particular application). In this way, the autonomous sensor continuously evolves its structure and self-calibrates (adapts its parameters). For example, if the model used is a FRB system it evolves by adding or removing fuzzy rules, selecting inputs and adapting parameters as described in Chapters 5 and 6 and as depicted in Figures 11.6 and 11.7 for the composition in the bottom of the distillation column and in Figures 11.9 and 11.10 for the polypropylene.

The comparison of the output by *AutoSense* with the values from the laboratory analysis (the target) is depicted in Figure 11.7 for the first subproblem.

The results in terms of root mean square error, RMSE and correlation, number of local submodels (respectively, rules) and inputs are also tabulated in Table 11.1 (using all input variables) and Table 11.2 (using online input variable selection as described in Section 5.4.1).

The accuracy of *AutoSense* for both cases is very good in both scenarios (using all inputs as well as using the online input variable selection). The structure (measured by the number of fuzzy rules, R) that was self-evolved is quite small and the generated fuzzy rules (Figure 11.8) are linguistic, that is, interpretable (Figure 11.10). These results demonstrate that *AutoSense* can work efficiently and self-calibrate even after a drastic change in the operating conditions by autonomous evolution.

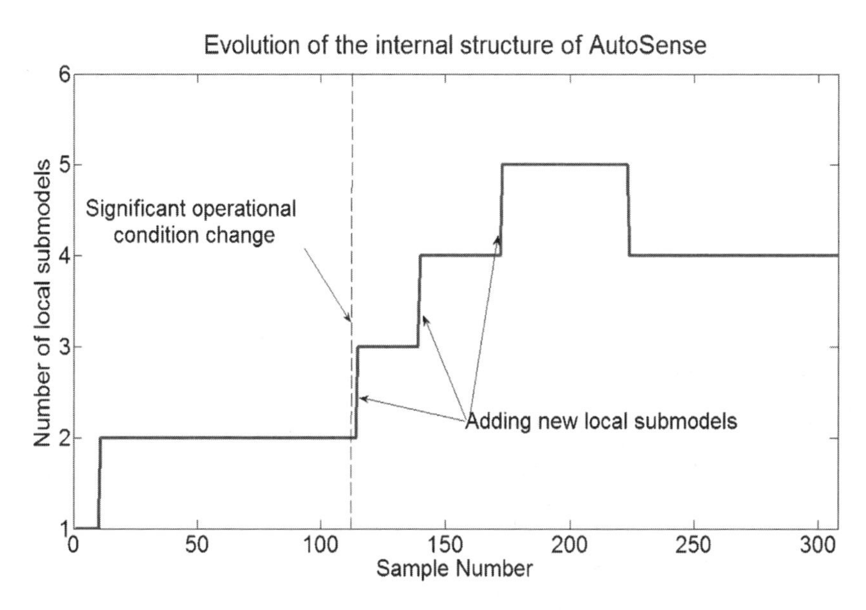

Figure 11.6 Evolving the structure of the model of *AutoSense* (first subproblem)

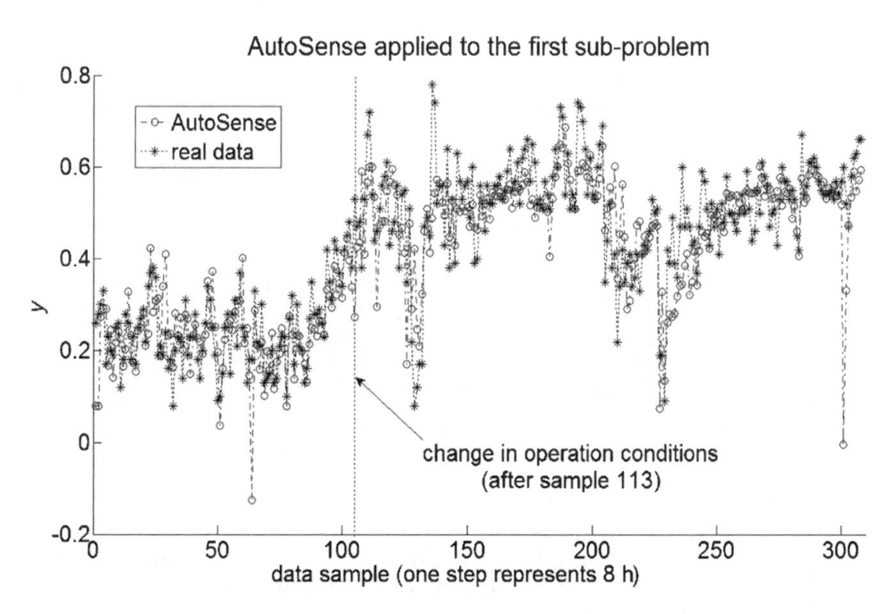

Figure 11.7 Output of *AutoSense* versus the 'true' values (measured using gas chromatograph)

Rule$_1$: ***IF*** *(*$x_1 \sim 24.6$*)* ***AND*** *(*$x_2 \sim 26.3$*)* ***THEN***
$(\bar{y} = -0.039 + \bar{x}_1 - 0.324\bar{x}_2)$

Rule$_2$: ***IF*** *(*$x_1 \sim 39.0$*)* ***AND*** *(*$x_2 \sim 43.5$*)* ***THEN***
$(\bar{y} = -0.615 + 4.77\bar{x}_1 - 0.340\bar{x}_2)$

Rule$_3$: ***IF*** *(*$x_1 \sim 46.2$*)* ***AND*** *(*$x_2 \sim 49.5$*)* ***THEN***
$(\bar{y} = -0.679 + 1.090\bar{x}_1 + 0.450\bar{x}_2)$

Rule$_4$: ***IF*** *(*$x_1 \sim 45.9$*)* ***AND*** *(*$x_2 \sim 49.9$*)* ***THEN***
$(\bar{y} = -1.340 + 5.570\bar{x}_1 - 3.320\bar{x}_2)$

Rule$_5$: ***IF*** *(*$x_1 \sim 36.2$*)* ***AND*** *(*$x_2 \sim 43.5$*)* ***THEN***
$(\bar{y} = -0.002 + 0.320\bar{x}_1 - 0.065\bar{x}_2)$

Rule$_6$: ***IF*** *(*$x_1 \sim 31.6$*)* ***AND*** *(*$x_2 \sim 38.7$*)* ***THEN***
$(\bar{y} = -0.007 + 0.366\bar{x}_1 - 0.129\bar{x}_2)$

Figure 11.8 *AutoSense* structure using FRB system as a framework for estimating polypropylene. $\sim$ means '*is around*'; $\bar{y}$, $\bar{x}$ denote the normalised inputs and outputs, respectively

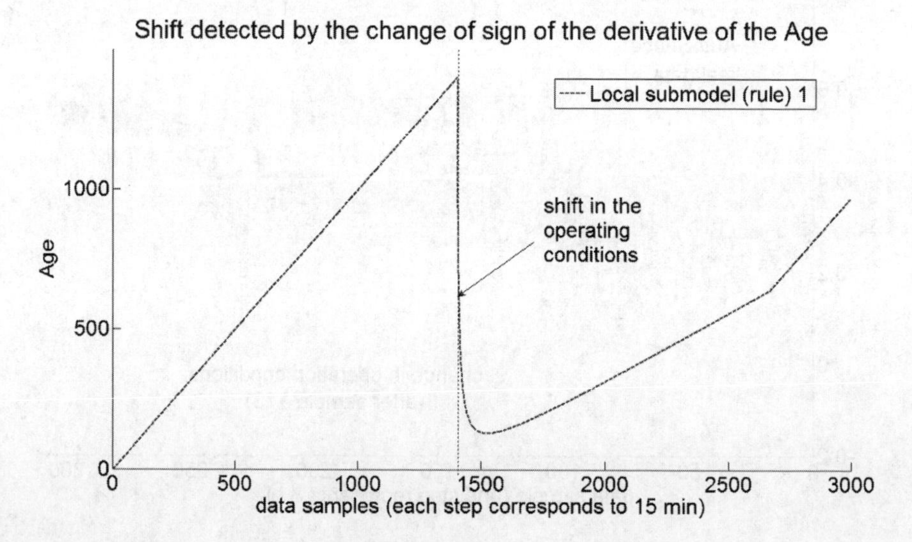

Figure 11.9 The *shift* in the data stream is visible around sample 1300 for the case study of propylene

Table 11.1 Results applying *AutoSense* to the chemical industry case studies ('composition' means 'composition in the bottom of the distillation tower' – the first subproblem)

	Composition	Propylene
RMSE	0.096	0.169
correlation	0.832	0.948
# submodels (rules)	3	4
n, number of inputs	47	23

Table 11.2 Results applying *AutoSense* to the same chemical industry case studies but including online input selection

	Composition	Propylene
RMSE	0.091	0.072
correlation	0.847	0.989
# submodels (rules)	3	6
n, number of inputs	4	2

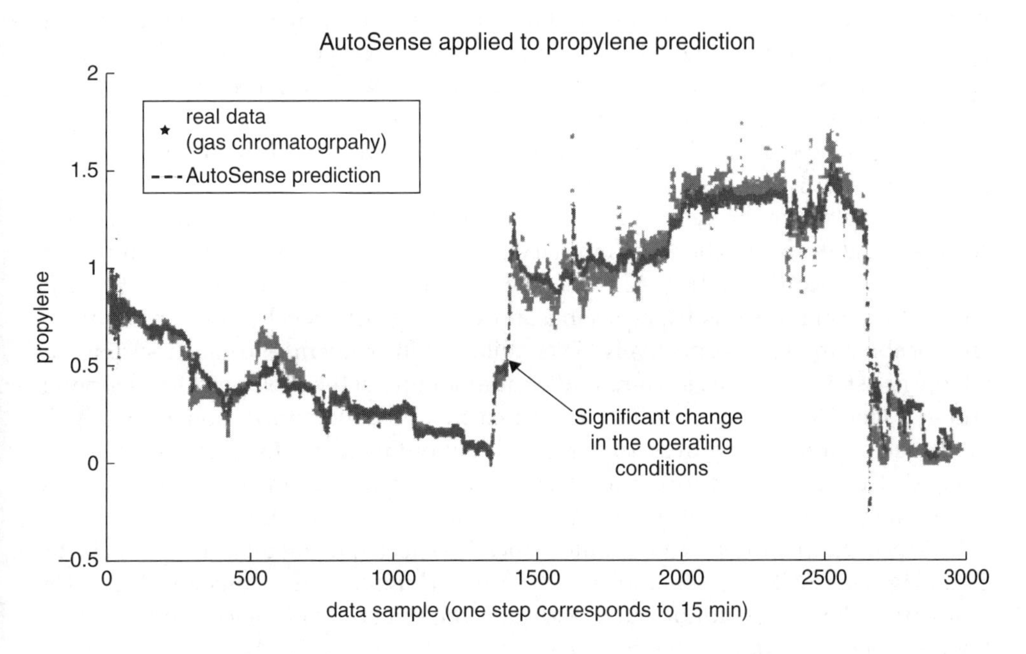

Figure 11.10 *AutoSense* for the propylene prediction

Moreover, Table 11.2 shows that the overall accuracy of *AutoSense* can be improved by using a more compact sensor model structure (fewer fuzzy rules, fewer inputs actually used). In addition, this structure simplification leads to a better interpretability. If we use AnYa-type FRB there will be an additional advantage of the lack of parameters in the antecedent part.

11.2.2 Drift and Shift Detection by Cluster Age Derivatives

The concept of so-called '*drift*' and '*shift*' in data streams (Widmer and Kubat, 1996) recently attracted a lot of attention in machine learning literature. *Drift* is the term used to denote the *gradual* evolution of the data stream over time. It refers to the smooth slide of the data distribution through the data/feature space from one region to another and occurs when there is a change of the distribution of the new data as compared to the old data. *Shift* refers to an abrupt change in the data density. *Drift* and *shift* provide a representation of the join time-data space dynamics, while data density is a representation in the data space with no explicit time.

The aim is to, first, detect the *drift* and *shift* and, then, act in accordance by changing/evolving the system (model/classifier/predictor/controller) structure taking into account the changed/evolved data density. For example, the *shift* is an indication that the system (model/classifier/predictor/controller) structure has to evolve adding a new local submodel to cover the new, unexplored region of the data space. *Drift* is related to a smooth change and can, therefore, be dealt with by replacing a local submodel with another, more relevant one. Traditional modelling/classification/prediction/filtering/control methods use a certain fixed structure and in the case of a *drift* or even *shift* in the data distribution they either represent the average (if the data affected by the *drift* and *shift* are part of the training) or lose their representativeness (otherwise).

Drift and *shift* can be caused by seasonal effects, wear and tear, contaminations, change of catalyser (in chemical industry), change of operating regimes, and so on (Angelov and Kordon, 2010). ALS are able to automatically detect and react to *drift* and *shift* in data streams by replacing and creating new focal points (prototypes) and local submodels, respectively. This ability is instrumental for such systems to handle nonstationary data streams and be autonomous self-developing, self-learning, and evolving. Moreover, it is important that this be done online. Some previously published approaches address *drift* and *shift*, but offline using, for example, the SVM method (Klinkenber and Joachims, 2000). This is, obviously, incompatible with the requirement for online real-time mode of operation demanded by the ALS.

In ALS the *drift* and *shift* can be automatically detected using the concept of cluster cloud *Age* (see Section 5.4.2) and its derivatives (Lughofer and Angelov, 2011). The gradient of the *Age* evolution curve indicates the dynamics of the data distribution. For example, a constant gradient (zero second derivative) of the *Age* curve indicates a stationary behaviour of assigning the data to a certain cluster/cloud/local submodel. A change of the gradient of the *Age* curve (nonzero second derivative) indicates a nonstationarity with the inflex point (point when the gradient changes its sign) indicating the *shift* of the data distribution and a change of the value (slope) but not the sign indicating a *drift*.

It is important also to distinguish between the *drift* and *shift* of the joint input–output data space and of the input or output data only. A good example of *drift* in the data pattern was observed with both case studies described in the previous section (around sample 113 and 1300, respectively, see Figures 11.6, 11.10.

The above paragraph describes how to detect *drift* and *shift*. Once this is detected, the model/classifier/predictor/filter/controller structure has to evolve to reflect that. If *drift* is detected the new focal point (prototype) and, respectively, local submodel that represents the new subarea of the data space to which the data stream *drifts* replaces the nearest of the previously existing ones from which the data distribution *drifts* away). If the *drift* is detected in the output variables only the model/classifier/predictor/filter/controller structure does not change but the learning of the local submodel (consequent) parameters increases the rate of forgetting (Lughofer and Angelov, 2011).

The reaction to *shift* is a substantial change of the model/classifier/predictor/filter/controller structure by adding a new focal point (prototype) and local submodel, respectively, to cover the new subregion of the data space.

11.2.3 Input Variables Selection

In these case studies the number of input variables was large (23 and 47, respectively). Many of them were highly correlated (which is often the case when working with industrial data). A thorough offline study was performed using genetic programming to identify the optimal subset of variables (Kordon *et al.*, 2003; Angelov, Kordon and

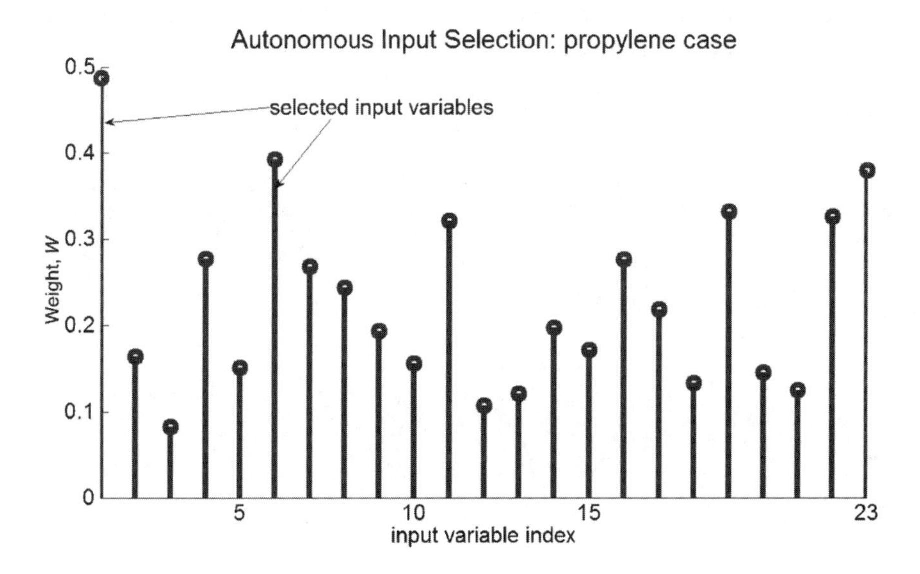

Figure 11.11 Autonomous online input variables selection

Zhou, 2008). It resulted in suggesting a significantly smaller subset of 7 and 2 input variables, respectively. Alternatively, applying *AutoSense* we found an even smaller subset of inputs for the chemical composition at the bottom of the distillation tower (namely, 4 inputs) and the same number of 2 inputs for the propylene. Moreover, the reduction of the inputs was done autonomously online as described in Section 5.4.1. Figure 11.11 illustrates the weights of the input variables that were used initially. The two inputs with the highest weight that are selected autonomously as described in Section 5.4.1 are indicated with an arrow.

11.3 Conclusions

In this chapter several special cases were considered from the chemical and petrochemical industries that have a lot in common, but also differ significantly. These include an oil refinery process and distillation unit, in particular, where *AutoSense* proved to be very useful for predicting the quality of various products such as kerosene, heavy naphtha, gasoil, and so on. It is also very useful in predicting the so-called inflamability (Abel) index.

AutoSense was also tested and applied to predict the composition in the bottom of a distillation tower in polymerisation processes. It was also applied to propylene modelling and prediction. The specific property of *AutoSense* to self-calibrate can save significant costs in the overall lifecycle and can, in addition, be used to automatically downselect a small subset of input variables needed.

In addition, the quality of the local submodels parameter *Age* (or more specifically, its second derivative) was proven to be directly linked to the so-called *shift* in the data pattern, which is an important research area in data stream mining and machine learning, but can also be linked to sudden changes in industrial processes, such as change of the operating mode, change of catalysers, and so on.

12

Autonomous Learning Systems in Mobile Robotics

The area of mobile robotics is, perhaps, the most natural area for implementation of ALS. By virtue of their role to sense, act, make decisions, plan and predict as humans but without direct human involvement robots are, in fact, the most obvious physical embodiment of ALS. In this chapter the application of ALS to laboratory-type mobile robots Pioneer 3DX (ActiveMedia, 2004) will be described based on the experiments made in Infolab21, Lancaster University by the author and his students.

12.1 The Mobile Robot Pioneer 3DX

In the experiments described in this chapter mobile robot Pioneer-3DX (ActiveMedia, 2004), supplied by ActiveMedia Ltd, Amehirst, USA was used, Figure 12.1. It is equipped with an onboard real-time controller called ARCOS, computer (Pentium III CPU, 256 MB RAM), an electronic-driven motor, a high-resolution real-time camera (Cannon), digital compass, sixteen sonar disks and bumper sensors, and a wireless connection. The laser scanner mounted on the robot has a resolution of 1^0 in spatial terms and detectable range up to 10 000 mm. All devices, including the motor and the sensors are controlled by the software loaded on the onboard computer through the onboard real-time controller ARCOS.

The application algorithms that realise higher-level '*behaviours*' such as 'wall following', 'novelty/landmarks detection', 'following the leader' are implemented using ARIA software suit. ARIA provides a wide range of foundation classes written in C/C++ language facilitating the control of the robot and onboard devices (ARIA, 2011) and runs on top of the ARCOS.

Autonomous Learning Systems: From Data Streams to Knowledge in Real-time, First Edition. Plamen Angelov.

Figure 12.1 Mobile robot Pioneer 3DX used in the experiments at Infolab21, Lancaster University, UK

From a software point of view, the robot can be seen as an autonomous agent with a five-layer architecture (see Figure 12.2).

At the lower layer are the devices such as 'hard' sensors (sonar, laser, electronic compass, motor drives, etc.). At a higher level is the embedded controller, ARCOS that acts as a server in a client-server mode with clients residing at the onboard computer at the higher layers. The embedded controller receives signals and data from the sensors at the lower level through interface and commands from the higher level through the software of ARCOS.

At the higher functional layer high-level programmes in ARIA that realise behaviours are implemented. The object-orientated class structure of ARIA simplifies the programming development and testing cycle time, enabling an easy access to the functionality of the mobile robot as a physical device such as to maintain the velocity of the wheels, adjust the robot headings, getting readings from the sensors, setting robot status, and so on (Angelov and Zhou, 2007).

At the top application layer the application programmes implement specific missions and tasks. This layer has the highest level of '*intelligence*' since it deals with problems such as automated reasoning, decision making, route planning, self-localisation and mapping (SLAM), object detection, identification, tracking/following, and so on.

12.2 Autonomous Classifier for Landmark Recognition

Mobile robots require for navigation in an unexplored environment a map (usually preloaded) and use of a global positioning system (e.g. the American GPS, the Russian GLONASS or the planned European Galileo). A very attractive alternative, especially in a dynamic environment (where maps and global positioning may fail), is to navigate using so-called *landmarks* and their relative position acquired automatically. It is superior to the so-called 'dead reckoning' (Kleeman, 1992) that is prone to drifting errors. The ability to autonomously extract landmarks and use them for localisation, navigation and routing may contribute to the survivability in cases when GPS/GLONASS is unavailable or unreliable. *Landmarks* can be defined as specific physical objects of the environment that stand out, are unusual and differ from the contextual background (Netto, 2006).

There is a significant difference if we compare indoor and outdoor environment; for example, GPS/GLONASS does not, in principle, work indoors, there are no trees and monuments indoors. Features of the architecture such as corners, corridors, doors can be used for navigation even if they do not fully comply with the definition of landmarks. It is significantly easier to build a map of an indoor environment due to the availability of long linear features, such as the walls and corridors.

A map of an outdoor environment can still be built but it is of a significantly more simplified and abstract nature linking the detected and identified landmarks with straight lines, see Figure 10.3. Such a simplified map can also include all points when a turn has taken place representing in this way a feasible (accessible) road (Sadeghi-Tehran *et al.*, 2012). In addition, outdoor landmarks are much more open to influence by the weather and environment that may change their appearance, including but not only in regard to the luminance. This includes seasonal and daily changes.

The ability to automatically detect and identify the landmarks and build a simple map based on them is clearly beneficial or even required in order to provide a level of autonomy and independence from the preloaded maps. Moreover, because of the autonomous character of the system, this should be done in an unsupervised and evolving manner. The reason for the latter requirement is that the number of landmarks is not known beforehand. The requirement for the learning to be unsupervised is obvious for an autonomous system. In addition, because of the constraints of miniaturisation of the size and/or to the payload this should also be computationally simpler and applicable in real time.

Finally, for the purpose of monitoring and in compliance with the famous Isaac Azimov's principles (that the robots/autonomous systems should never threaten humans), the landmarks and simple maps generated autonomously should be transparent and interpretable. The methods and algorithms described further in this section address all of the above requirements. They are based on RDE, *AutoCluster* and *AutoClassify0* described earlier and perform joint landmark detection, identification and classifier design.

Alternative approaches for unsupervised learning such as SOM (Kohonen, 1982, 1984, 1995) and evolving and self-organising neural network such as growing cell structures, adaptive resonance theory (ART) mapping (Carpenter and Grossberg, 2003), dynamically evolving neurofuzzy inference systems (Kasabov and Song, 2002), resource allocation networks (Plat, 1991) do not take into account the data density and are prone to generate too many clusters that are hard to use as landmarks.

All these approaches are not prototype-based; the cluster centre in all of them is the mean/average and/or is a result of an adaptation, thus, being an abstraction that may be located in an infeasible point of the data space. In addition, the new data sample is compared to the cluster centers only, not to ***all*** previous data because the real-time nature precludes memorising the data history. In this way, important information is usually lost.

In contrast, the RDE takes into account ***all*** previous data samples; *AutoCluster* and the related *AutoClassify0* are prototype-based. They are fully unsupervised in the sense that not only the labels/outputs but also the number of prototypes are not predefined but are determined based on the data density. In *AutoClassify0*, in addition, class labels (landmarks ID) are assigned automatically.

12.2.1 Corner Detection and Simple Mapping of an Indoor Environment through Wall Following

The first experiment involves a mobile robot Pioneer 3DX moving autonomously in an office environment (ActiveMedia, 2004). The motion is considered in a 2D plane (the floor) with corners as main 'landmarks'. In order to move autonomously, the mobile robot needs a map with 'way points'. In an autonomous scenario instead of this being preloaded or provided by a human there should be an ability to generate it automatically even in a completely unknown environment (Zhou and Angelov, 2007).

Landmarks can be extracted autonomously by the robot implementing an ALS in its embedded computer collecting the data by its sensors (in this case, sonar and odometer sensors were used only, but radar can also be used for more precise results and/or camera for visual landmarks as described in the next section) while travelling applying so-called *wall following*. *Wall following* itself is a relatively simple control problem aiming to move in parallel to the walls, and in this way, to go round the room exploring and identifying the corners that can, generally, be of two types – convex and concave (Zhou and Angelov, 2006).

Based on the corners, their type and the distance between them one can build a simple map of the room and use it later, transmit it to a centre or another robot if working in a team/swarm and even arrange rendezvous (Zhou and Angelov, 2006; Angelov, 2006). Such a scenario does not require a direct human involvement (except, possibly, the ability to monitor or abort the mission), the use of GPS/GLONASS, preloaded maps, plans and controllers design.

The odometer (inertial sensor) provides information about the relative coordinates (the start point has always coordinates (0,0) in a 2D plane) from which the distance travelled between turnings can be extracted. At the top layer (see Figure 12.2), an ALS was implemented that realises RDE and evolving clustering.

A similar, but much simpler experiment was realised by Nehmzow, Smithers and Hallam (1991) who used offline supervised learning with a fixed stricture model (predefined number of corners, or, more generally, landmarks). Zhou and Angelov (2006, 2007) used RDE and eClusteirng in experiments performed in B69 office of Infolab21, Lancaster University, UK. This particular office has eight rectangle corners (six concave and two convex), see Figure 12.3.

The aim of the experiment was to uniquely identify the corners of the room (used as simplified landmarks in this experiment) and, based on them and their position, to

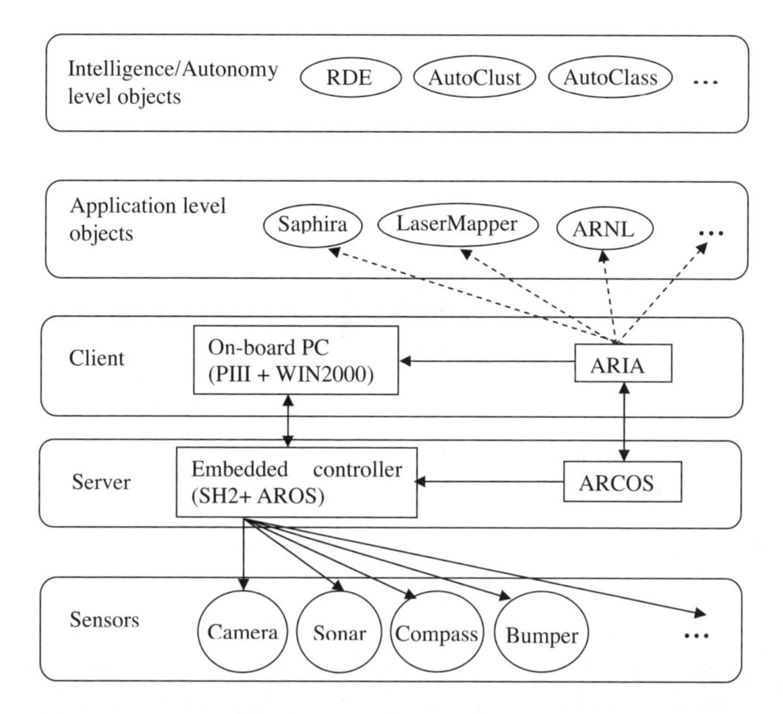

Figure 12.2 The architecture of the software on Pioneer 3DX

build a simplified map (which can be used for a navigation or arranging rendezvous). Two input variables were measured only, namely

a. φ_i – rotation of the robot when it steers around the corner; and
b. d_i – distance between the robot and the previous corner.

where

$i = [1, K]$
K is the number of corners, which is ***not*** predefined).

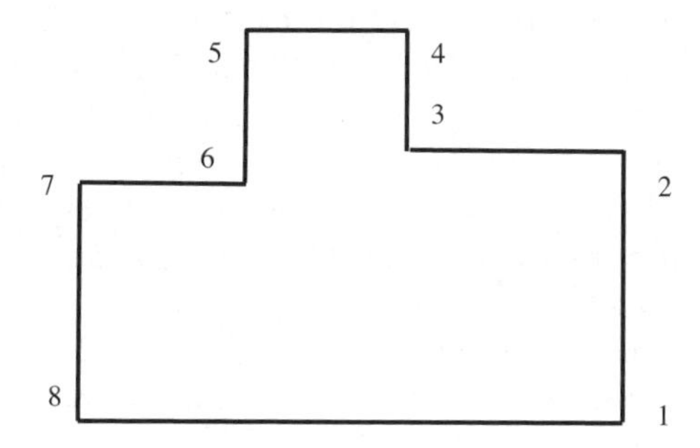

Figure 12.3 A sketch of the office B69, Infolab21, Lancaster University, UK

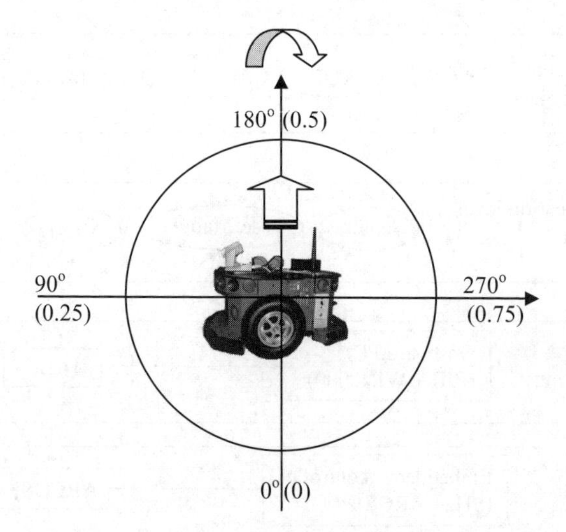

Figure 12.4 Rotation of the Pioneer robot – schematic representation

If we compare the experiment settings, they differ from the one described in (Nehmzow, Smithers and Hallam, 1991) primarily by the fact that K is not predefined and no pretraining by a 'teacher' (human being) is used.

The rotation is measured clockwise in degrees (or radians) and is normalised by 360° (or 2π, if use radians) starting with 0 representing the straight back (reverse) direction from the robot, Figure 12.4. For example, rotation $\varphi = 0.5$ corresponds to heading straight forward; rotation $\varphi = 0.25$ corresponds to turning left at a right angle; rotation $\varphi = 0.75$ corresponds to turning right.

The distance, δ_i is measured in meters by the rear sonar device (note that only one out of the available sixteen sonars is used in this experiment). The values are normalised to conform to the range [0;1] normalising with the value of the range, which is 10 m (ActiveMedia, 2004).

The data that are collected at each time instant, k with a frequency of 1 Hz are in the form of a vector, $x_k = [\varphi_k, \delta_k]$. From a methodological point of view, the novelty/anomaly detection method (Section 2.6) based on RDE (Section 2.5) is applied or, alternatively, an *inverted* version of the eClustering as reported by Zhou and Angelov (2006, 2007) or an autonomous classifier of type *AutoClassify0*. '*Inverted*' here means that a new cluster is formed around data points with density values lower than the values of the already existing ones (which is precisely the opposite to the eClustering method, where new clusters are formed around points with values higher than the values of the already existing ones).

From the software implementation point of view, the classes *ArRobot* and *ArSonarDevice* from the ARIA library are used to control the velocity of the wheels of the robot and the sonar sensors, respectively. The *wall-following* behaviour is implemented as an algorithm defined in class *RecognitionApp*.

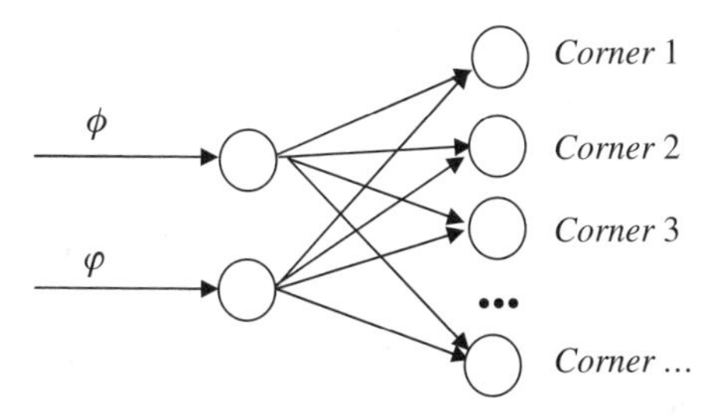

Figure 12.5 ALS as a NN that self-evolves

Note that the number of the corners is ***not*** predefined and the algorithm is applicable to an arbitrary/unexplored room. Each corner is a focal point (a prototype) around which fuzzy rules (or respective neurons or probabilistic data distributions) can be defined, Figure 12.5:

$$R_1 : IF \left(\begin{bmatrix} \varphi \\ \delta \end{bmatrix} is \sim \begin{bmatrix} \varphi_i^* \\ \delta_i^* \end{bmatrix} \right) \quad (12.1)$$
$$THEN(Corner\,1)$$

This autonomous classifier can be used to associate future data with one of the previously encountered corners or to evolve the autonomous classifier (*AutoClassify0*) further by adding a new corner. Moreover, this approach can be used for SLAM and mapping of an unknown environment because the distance and relative position of the corners is discovered autonomously rather than being provided (even the number of corners does not need to be provided).

In the experiments reported by Zhou and Angelov (2006, 2007) seven out of the eight existing corners were autonomously identified. Only one corner was misclassified due to the noisy real data. These results (Zhou and Angelov, 2006) were superior in comparison to the results reported by Nehmzow, Smithers and Hallam (1991) and, in addition, there was no pretraining (the autonomous clustering and classification based on RDE started '*from scratch*').

12.2.2 Outdoor Landmark Detection Based on Visual Input Information

In a separate set of experiments performed outdoors in the campus of Lancaster University the aim was to detect and identify autonomously landmarks extracted from video input from a robot camera mounted on Pioneer 3DX. The video signal used as input was processed online in real time, while the robot travelled within the campus. This scenario posed additional challenges in the sense that the mobile robot had to control concurrently in real time several processes, namely:

i. control and interface with the camera acting as a sensor in this scenario;
ii. control of the wheels and motion of the robot; and
iii. processing of the video stream.

For building a map in order to identify the relative location at the time instances when a new landmark is detected an odometer was used in addition to the camera. Different features can be extracted from the imagery, but most of them may require computationally expensive processing (e.g. edge detection combined with object identification, shape and size of objects etc.). A computationally efficient alternative is to use the colours (red, green and blue, RGB) or the transformations such as hue, saturation and value of brightness (HUV) which are readily extracted from the images (Zhou and Angelov, 2007).

As the robot moves in a previously unseen and unknown environment it generates a video stream using the onboard camera. With the proposed algorithm based on RDE, *AutoCluster* and *AutoClassify0* the image frames are processed in real time. The current frame is only used and discarded (only accumulated statistical information about the distribution of the colour information is kept in the memory). The image frames that have significantly lower density and are distinct are detected automatically and declared as landmarks. They are automatically assigned an ID (label) and a fuzzy rule-based classifier of the following form has been autonomously evolved:

$$\begin{aligned} Rule_i &: IF\ (x_k \sim x_i^*) \\ &THEN (x_k\ is\ LM_i) \end{aligned} \tag{12.2}$$

where

$i = [1, R]$;
R is the number of fuzzy rules; the consequent LM_i is the ith landmark;
x_k – k-th image frame;
x_i^* is the prototype of the ith rule antecedent.

Note, that due to the fuzzy nature of the rules the similarity between images is measured by a degree (they do not need to be exactly the same), which contributes to the flexibility and robustness in recognition of previously seen scenes that may differ slightly in terms of appearance, luminance, angle of view and so on.

The algorithm starts with grabbing the bitmap image frame with size $H \times V$ pixels and segmenting it into a grid of *bins* (Zhou and Angelov, 2007) – smaller subareas of the image frame of $m \times n$ pixels that may contain objects that will be used as landmarks. In the experiment that was conducted at Lancaster University image frames of size 640×480 pixels separated into twelve 160×160 pixel bins were used, as illustrated in Figure 12.6.

In each bin, the mean value of the colour intensity of all pixels is calculated for each colour (Red, Green, and Blue):

$$\mu_{ij}^{R} = \frac{1}{mn} \sum_{i=1}^{m} \sum_{j=1}^{n} I_{ij}^{R} \tag{12.3}$$

Figure 12.6 Image (with size 640 × 480 pixels) segmented into12 bins with size 160 × 160 pixels

where

I_{ij}^{R} denotes the intensity of the Red of the ith column; j^{th} row of the bin;

μ_{ij}^{R} denotes the mean value of the Red in the bin formed by the ith vertical and j^{th} horizontal of the image ($i = [1, 3]; j = [1, 4]$).

The inputs/features of the autonomously evolving classifier are formed by the three colours of each bin, which make thirty six features, Figure 12.7.

The experiment was conducted in the campus of Lancaster University, UK. This experiment, described in more detail by Angelov and Zhou (2007), took about 6 minutes during which a real-time video stream was generated by the camera mounted on the robot. The camera took shots at a rate of 25 fps, but the frames used by the algorithm were processed by the embedded computer at a rate 11.6 fps.

The algorithm accumulates recursively the similarity (of the current image frame with all of the previous image frames) represented through the density calculated using RDE as described in Section 2.5. Let us recall that the landmarks are defined as 'physical objects of the environment that *stand out, are unusual and differ* from the contextual background'. Therefore, the image frames that cause a drop of the density represent a landmark (which is in other contexts, an anomaly, or outlier). The landmarks are kept in the memory and each new landmark identified based on the drop of the density is also compared with the previously encountered landmarks to detect if it was already seen (which would mean that the mobile robot has already seen this landmark and had been there – closing the loop). When linking this with the relative location data from the odometer of the mobile robot a simple map can be derived autonomously and without any intervention of humans or use of GPS/GLONASS.

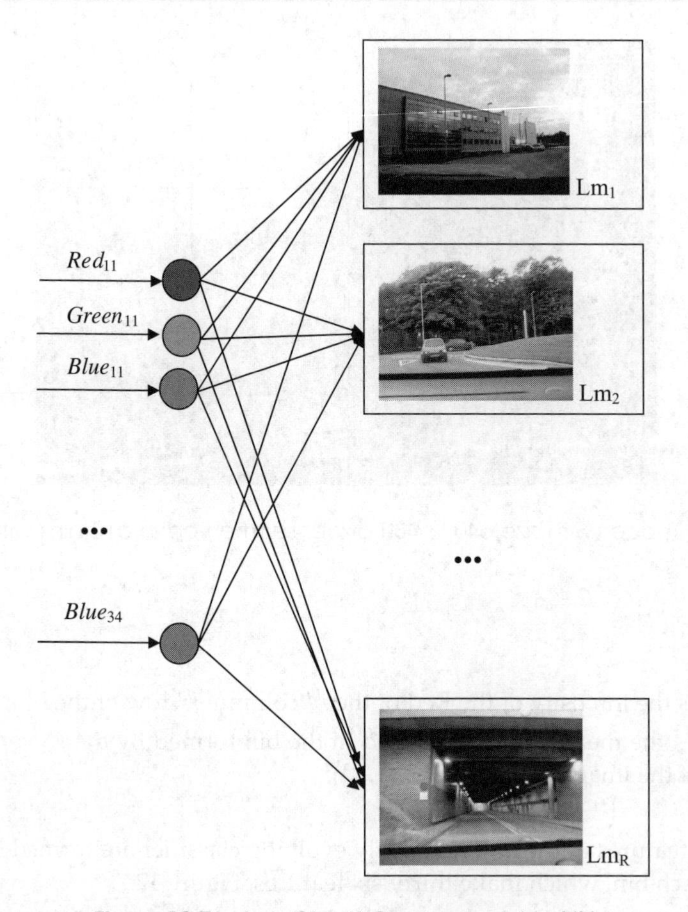

Figure 12.7 *AutoClassify0* as an evolving NN

This experiment illustrates the capability of *AutoClassify0* to autonomously self-develop its classification hypersurface, (rule-based) structure and simultaneously to classify the incoming video stream into:

i. landmarks; and
ii. images that are rather routine.

In this way, higher-level knowledge (in the form of simple map, a set of images representing landmarks visited and rules, based on which the decision is based) can be extracted without any human intervention. Such higher-level information is human-intelligible and understandable and can be transmitted to a human or another mobile robot (in a collaborative scenario). A very important, innovative and useful feature is the open-ended form of *AutoClassify*, which means that the number of the landmarks does not need to be predefined, but is rather extracted from the environment. This flexibility is very important for the autonomous systems because it is unrealistic to know in advance the environment. In addition, because of the recursive and noniterative nature, this method is also computationally very light, which is also very important for any real-time application.

12.2.3 VideoDiaries

In this subsection another application will be described which can automatically acquire a sequence of visual landmarks – places visited during the day with time and location assigned to them that may form an improvised VideoDiary. It can be realised as a *smart* phone app (Angelov, Andreu and Vong, 2012) or as a *smart* wearable microcamcorder fixed to the head of the patient such as the one produced by Microsoft's SenseCam (Berry *et al.*, 2007).

Nowadays, *smart* phones are involved deeper in our everyday life and are, perhaps, the most popular device with elements of *intelligence*. They are handheld and/or wearable, equipped with a range of sensors, processing capability, memory, *wifi* and GSM as well as cameras and inertial sensors, often with GPS/GLONASS and so on. The VideoDiary app on a *smart* phone can be an attractive application for young users.

The application on a wearable microcamcorder can be very useful for patients diagnosed with neurodegenerative disease. Personal experiences can be stored as pictures, in an autobiographical collection of "flash" memories. Previous research has shown that the use of such devices for memory-impaired individuals, improves successfully the recall of personal experiences (Hodges *et al.*, 2006).

This increases the flexibility of the programmers who do not necessarily need to develop special cases for each type of streaming sources.

An application was developed in the form of a software suite with several "views" and "screens". The main class or widget is responsible for handling the object menu, between the FrameGrabber screen, density graph and album diary.

The user can navigate through the following screens of the application:

- Life Content Window: This is the default view to be displayed when a video source is selected. The current frame, the last captured landmark and some informative data as shown in Figure 12.8.
- Diagram Window: This view shows (Figure 12.10) all the captured landmarks during a single record and visualises the values of the density, mean and standard deviation of the density over a period of time. The view is updated frame by frame.
- GPS Mapper: This view is only available when GPS data is being recorded. It shows the captured landmarks over the trajectory displayed in red. The current user's position is shown as a blue circle, Figure 12.9.
- Diary Window: This view implements a simple interface for reviewing captured landmarks. Users can view the landmarks by date or by video files and scroll through the view.
- Application Dialogues: The application dialogues include the main window and auxiliary windows used to select video sources, to change Landmark Recognition's parameters, view and to enter the diary viewer.

Figure 12.8 Life Content Window (from Angelov, Andreu and Vung © IEEE 2012)

- Settings Dialogue: From this screen users may change the camera dimensions to tune the number of pixels retrieved. Besides, users can define the number of bins, they request to use. For novice users "simple" type's settings are provided. If the bin dimension is high, smaller landmarks can be isolated in a bin and *vice versa*.

12.2.4 Collaborative Scenario

One application of the collaborative scenario is related to multiple (possibly, independent) agents working on information related to the same objects or people. It is an interesting dialectic phenomenon that, on the one hand, the algorithms that were described earlier in Part II of this book and on which ALS are based on are order

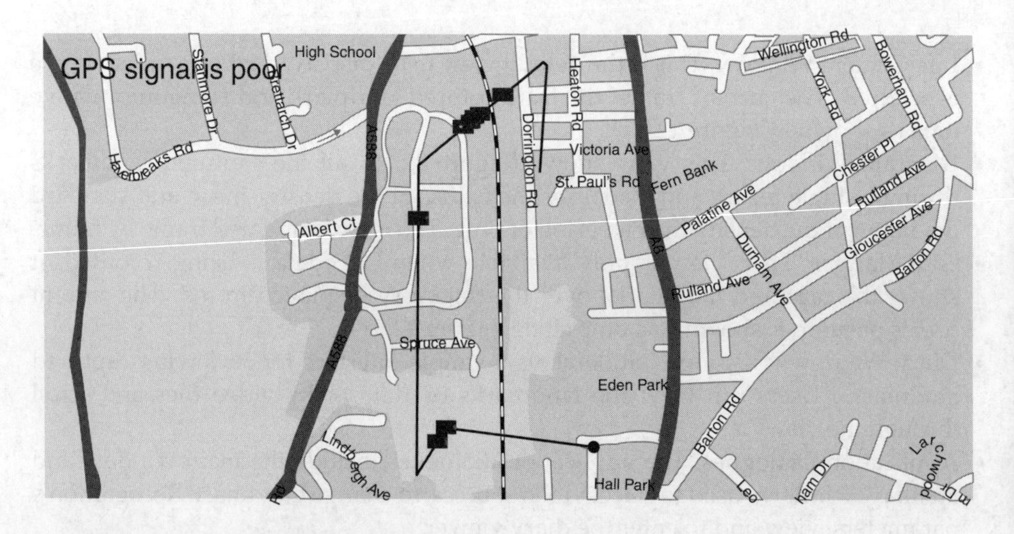

Figure 12.9 GPS Mapper View (form Angelov, Andreu and Vung © IEEE 2012)

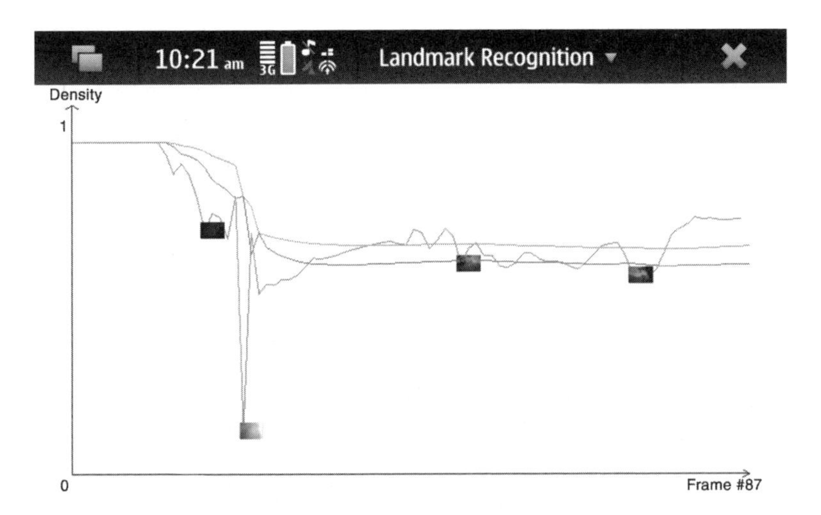

Figure 12.10 Example of the application (courtesy of Mr. Javier Andreu)

dependent (because they are one pass, recursive, online and work on data streams), although they are evolving and tolerate uncertainties optionally using fuzzy logic in their inference. At the same time, these algorithms always provide the ***same result*** if applied to the same data if we keep the same order. Therefore, they combine the uncertainty of the flexible and open structure, tolerance of deviations from the prototypes with the determinism of the statistical learning principles based on the data density distributions. In this sense, they are very much like the probabilistic models, but unlike them they consider the uncertainty not as a chance or belief but as a representation of the mismatch between the complexity of the real-world problems, processes and data streams and the simplicity of the 'convenient' theoretical representations such as linear models, normal distributions and so on.

One very interesting and useful application of the fact that these algorithms have a strong element of determinism and predictability of their behaviour is the collaborative scenario, which is described in this subsection. The main idea is that a given data stream can be processed either by one ALS or a multitude of ALS can process the same data stream (with the order of data kept unchanged!) piece by piece, Figure 12.11 (see also Chapter 10).

This very useful property was proven by Angelov (2006) and demonstrated in the project '*Multisource Intelligence: STAKE: Real-time Spatio-Temporal Analysis and Knowledge Extraction through Evolving Clustering*' (2011) funded by Centre for Defence Enterprise, UK Ministry of Defence on the examples of phone call data. It is very interesting that the result of applying any of the two schemes described above is guaranteed to be ***exactly the same***. At the same time, the benefits of the collaborative way of processing the data stream are that in this way:

a. the amount of data processed by each ALS can be significantly reduced (the so-called principle *divide et impera* can be used);

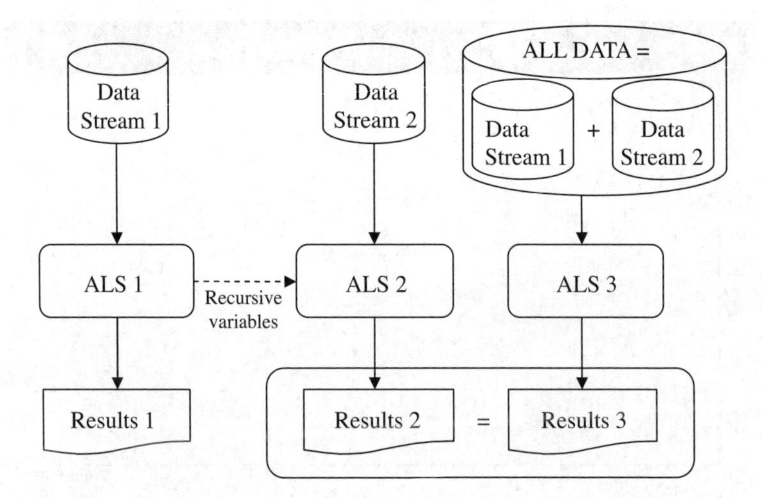

Figure 12.11 Collaborative scenario – schematic diagram

b. the amount of data transmitted between the ALS and, therefore, the bandwidth required is minute as compared to the amount of raw data processed (equal to the amount of the focal points/prototypes identified so far plus the accumulated statistics – densities, means, covariances, radii, *Ages*, supports, utilities);
c. each ALS has access only to a part of the data (in some scenarios this may be desirable and important for higher secrecy);
d. survivability and mission success of the team can increase significantly because if, for example, one ALS disappears or stops functioning for some reason it is enough that the focal points/prototypes and the accumulated statistic as described in item b) is transmitted to the other ALS to get the same result as if all ALS were intact (the dashed line in Figure 12.11).

On the right-hand side of Figure 12.11, we can observe that only ALS3 receives ***all*** the data and it processes this data using an ALS, as we have explained before. On the other side of the figure, one can see that the same process is done by two different ALSs (ALS1 and ALS2). In this case, the input to the ALS1 is the part of the Data Stream 1. This ALS processes this data, and transmits to ALS2 the focal points/prototypes identified and the accumulated statistics ***only***. ALS2 then starts to (continues if consider all data) receive the Data Stream 2. It is important to stress that, although ALS2 has not seen ***all*** the raw data (not seen Data Stream 1, to be more precise), the result of the processing is ***exactly the same*** as the one that can be obtained by the ALS3 that has seen all the data. One can think of ALS2 as being initialised by ALS1.

The list of variables that has to be transferred from ALS1 to ALS2 includes:

- vector of current focal points/prototypes, x^* (size $R \times n$) and their densities, D (size $R \times 1$);

- vector of mean values, μ (size $R \times n$);
- the scalar product, Σ (size $n \times n$);
- *Age*; utility, η; support, N_i- all with size $R \times 1$;
- radius, r of clusters (size $R \times n$), and;
- number of samples processed, k.

Another application scenario for the collaborative ALS is the team of uninhabited vehicles (UxV). For example, interesting missions may include the same SLAM problem considered above, but performed by a team of mobile robots rather than by single robot. This may offer additional operational capabilities such as arranging *rendezvous* at previously unspecified time and location (during the mission) without using a preloaded map or GPS/GLONASS and, in this manner, moving significantly forward in the direction of achieving a higher level of autonomy.

A particular mission involving multiple UxVs is, for example, the problem of formation control. It has been investigated in a variety of applications for unmanned aerial vehicles, UAVs (Ren and Beard, 2003), autonomous underwater vehicles (Stilwell and Bishop, 2000; Azimi-Sadjadi *et al.*, 2002; Kaknakakis *et al.*, 2004; Carline *et al.*, 2005), AUVs, unmanned ground-based vehicles (UGVs) (Desai, Ostrowski and Kumar, 2001), and so on. Overall, the approaches to formation control of mobile robots can be grouped into the following three main categories:

a. behaviour based;
b. virtual structure; and
c. leader following.

The last one of these methods was implemented as an ALS by Sadeghi-Tehran *et al.* (2010).

12.3 Autonomous Leader Follower

The *AutoControl* method described in Chapter 9 was applied to the *'Leader-follower'* problem in the framework of a UK Ministry of Defence, Centre for Defence Enterprise funded project *'Assisted Carriage: Intelligent Leader Follower'* (2009). The problem is illustrated in a very simplistic manner in Figures 12.12 and 12.13.

The *'Leader'* ('L') and the *follower* ('F') are defined by (x_L, y_L, θ_L) (v_L, ω_L) and (x^F, y^F, θ^F) (v^F, ω^F), respectively. The relative distance between the *leader* and the *follower*, δ_{LF} as well as the angle between the headings of the *'Leader'* and the *follower* θ_{LF} can be expressed from simple geometrical considerations:

$$\delta^{LF} = \sqrt{(x^L - x^F)^2 + (y^L - y^F)^2} \tag{12.4a}$$

$$\theta^{LF} = \arctan\left(\frac{y^L - y^F}{x^L - x^F}\right) \tag{12.4b}$$

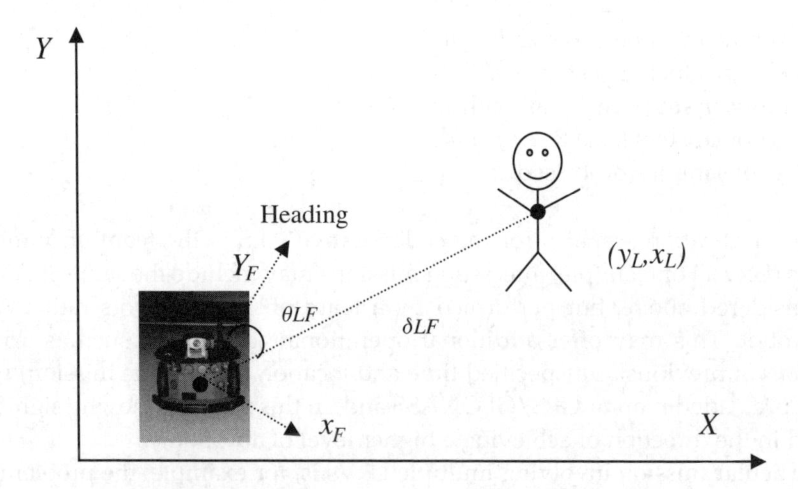

Figure 12.12 Leader Follower – a schematic diagram

Figure 12.13 Leader Follower – an illustration of the outdoor experiment that was performed in the campus of Lancaster University, UK (in the picture is Mr. P. Sadeghi-Tehran, a PhD student of the author). The video is available on YouTube (http://www.youtube.com/watch?v=4OEOgLSnoak)

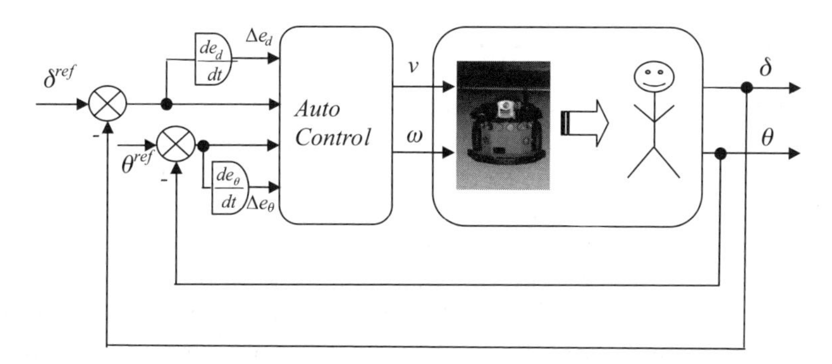

Figure 12.14 A schematic diagram of the autonomous learning controller *AutoControl* for the '*Leader-follower*' task

The diagram of the autonomous learning MIMO controller (*AutoControl*) with four inputs (error and derivative of the error for δ and θ respectively) and two outputs (linear and angular velocities) is designed to maintain the reference values for the distance ($\delta^{\text{ref}} = 0.5$ m) and bearing ($\theta^{\text{ref}} = 0°$), see Figure 12.14.

The inputs and outputs are defined as follows:

$$e_k^d = \delta_k^{ref} - \delta_k^{LF} \tag{12.5}$$

$$\Delta e_k^d = e_k^d - e_{k-1}^d \tag{12.6}$$

$$e_k^\theta = \theta_k^{ref} - \theta_k^{LF} \tag{12.7}$$

$$\Delta e_k^\theta = e_k^\theta - e_{k-1}^\theta \tag{12.8}$$

$$v_{k+1} = AutoControl\left(e_k^d, \Delta e_k^d\right) \tag{12.9}$$

$$\omega_{k+1} = AutoControl\left(e_k^\theta, \Delta e_k^\theta\right) \tag{12.10}$$

AutoControl develops 'from scratch' a set of local submodels of the type:

$$IF \left(\begin{bmatrix} e^d \\ \Delta e^d \\ e^\theta \\ \Delta e^\theta \end{bmatrix} \sim \begin{bmatrix} e^{d*} \\ \Delta e^{d^*} \\ e^{\theta^*} \\ \Delta e^{\theta^*} \end{bmatrix} \right) \quad THEN \left(\begin{bmatrix} v \\ \omega \end{bmatrix} = C\lambda \begin{bmatrix} v^* \\ \omega^* \end{bmatrix} \right) \tag{12.11}$$

designed as described in Chapter 9.

The use of *AutoControl* helps to solve a number of problems. The plant/object of control in this case is highly uncertain (the movements of the '*Leader*' are hard to predict and are characterised by randomness, nonstationarity and nonlinearity). Applying a traditional stochastic, neural network or fuzzy logic controller would have problems with the above-mentioned characteristics and, in addition, would have a complex and nontransparent structure. For example, even if we apply a FLC with Mamdani-type fuzzy sets (the most transparent and human intelligible type of controllers of the above list) there will be a large number of fuzzy rules (of the order of tens or even hundreds) and an even bigger number of parameters (of the order of hundreds).

In order to cope with the nonlinearity, let alone the nonstationarity, frequent adaptation and (offline) redesign is needed that requires a lot of training data with the same statistical characteristics as the data from the expected exploitation process. For example, Sadeghi *et al.* (2010) reported that applying a traditional Mamdani-type FLC required 49 fuzzy rules for this task. In addition, such a controller requires 'tuning' based on predefined training data that does not necessarily guarantee same performance in another environment making such a solution inflexible.

Alternatively, the *AutoControl* described in Chapter 9 requires for the same problem only nine focal points and fuzzy rules respectively (Sadeghi-Tehran *et al.*, 2010). Additionally, *AutoControl* can start working *'from scratch'* and learns and evolves during the process of control with the real (not specially preselected) data used as training on a sample by sample basis in the same way as adaptive control and estimation works.

12.4 Results Analysis

In this chapter different applications of ALS to mobile robotics were presented. Experimentations took place with laboratory-type Pioneer 3DX mobile robots at Infolab21, Lancaster University, UK in the last five years or so by the students of the author. Using *AutoCluster*, *AutoClassify* and *AutoControl* completely unsupervised learning of the environment in terms of identifying landmarks, mapping and navigation can be achieved. More details on this and other applications related to mobile robotics can be found in the journal and conference papers by the author and his students listed in the references. Here they are presented with the prime aim to illustrate the methodology presented in Part II of this book.

13

Autonomous Novelty Detection and Object Tracking in Video Streams

13.1 Problem Definition

In machine learning the problem of autonomous novelty detection is not new and has been studied in relation to fault detection and video analytics extensively. It aims to identify a new or previously unknown data item in the data stream or object in a video. It plays a pivotal role in a range of applications such as computer vision and robotics, surveillance and security, machine health monitoring, medical imaging, human–computer interaction, and so on.

In this chapter this problem will be considered from the point of view of its application to autonomous video-analytics. Same principles are behind the pioneering results on autonomous real-time anomaly detection and flight data analysis (FDA) in aviation which resulted in an EU project SVETLANA (http://www.svetlanaproject.eu/). This has an increasing importance nowadays when there is a huge and growing amount of video streams produced (for example, in UK there are over 4 million CCTV cameras installed, which makes one camera for every 14 people in the country (Daily Mail, 2009)) that require real-time analysis. The main aim of such an autonomous system (Hampapur, 2005) would be to detect, identify/classify, and track anomalous movements behaviour and activities in the area being observed/monitored.

The data stream can take the form of a video (image frames) from a digital camera, electro-optical (EO), infrared (IR) or other source, for example signature aperture radar (SAR) or passive millimetre wave (mmW). Video stream is usually taken with the frequency/rate of 25 to 30 frames per second (fps) to make the replacement invisible for the human eye. Most of the existing methods for processing video input are offline.

Autonomous Learning Systems: From Data Streams to Knowledge in Real-time, First Edition. Plamen Angelov.

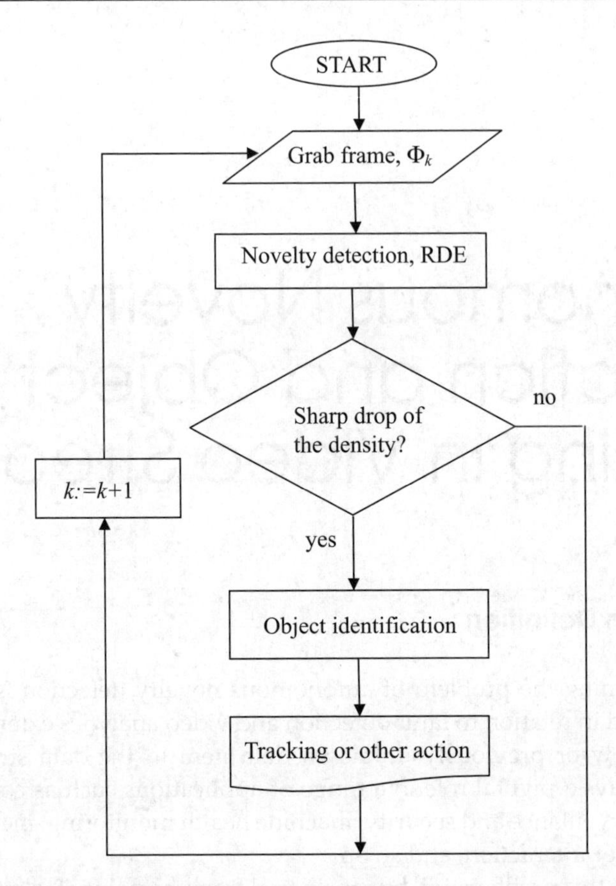

Figure 13.1 Schematic representation of real-time video processing (the last phase, for example, may be trajectory classification or other action)

The method, which is a step towards autonomous video analytics, was recently proposed by Angelov, Ramezani and Zhou (2008). It makes use of RDE and can be summarised by Figure 13.1.

13.2 Background Subtraction and KDE for Detecting Visual Novelties

13.2.1 Background Subtraction Method

Background subtraction (BS) is one of the most widely used methods for novelty detection that is based on modelling the background (bg) of the scene by statistical learning and subtracting this form the current image frame. In this manner, the foreground (fg) or the novel, unexpected object passing or appearing on the scene is being separated and identified. Such models have to be robust to noise, illumination changes and occlusion, representative and, at the same time, sensitive.

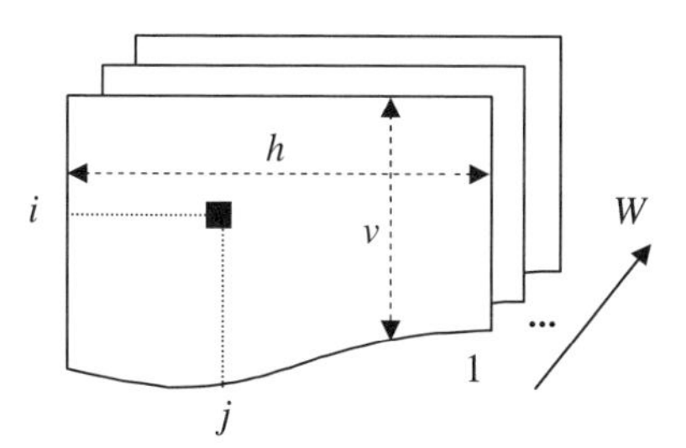

Figure 13.2 A window of *W* frames used in KDE approach, *h* denotes the number of pixels in the horizontal and *v* – the number of pixels in the vertical direction; the $(i,j)^{th}$ pixel is indicated by a black square

An intuitive technique is to compare the pixels of the current image frame with the pixels of previous frames having the same position (e.g. from a window of frames with a size W, as it is shown in Figure 13.2) in terms of some features, for example, colour (R,G,B) or hue saturation and value (H,S,V). The size of the window is usually 40 to 70 frames.

In this way, a separate model is built and updated for each pixel. The model estimates the density of a particular, $(i,j)^{th}$ pixel in comparison with the pixels in the same position from previous, *W* frames of the window. If the density has a low value (below a prespecified threshold) it is assumed to be a foreground. Usually, Gaussian normal distributions are assumed as a model of 'normality' (the background) and colours as features. The assumption is that the colour of the moving objects is different from the background.

The robustness of the method is required in order to distinguish between noise that results from movements of the tree leaves, branches and bushes, variation of the luminance during the day and so on and the genuine appearance of a new object on the scene. The sensitivity is required in order to minimise false negatives (FN), that is, to avoid missing a genuine new object that actually appears on the scene and to tackle occlusion and use of camouflage.

13.2.2 Challenges

There are a number of challenges that an ideal approach has to address. These are briefly outlined below.

13.2.2.1 Illumination Changes

It is well known that the illumination varies during the day and this affects the appearance of the background. The brightness may also be affected by the external sources of lights such as street or car headlights, lamps, sun or the moon, and so on. Clouds also affect the brightness as well as the corridors or mountains. Such changes may lead to the background model becoming irrelevant or incorrect or even to be

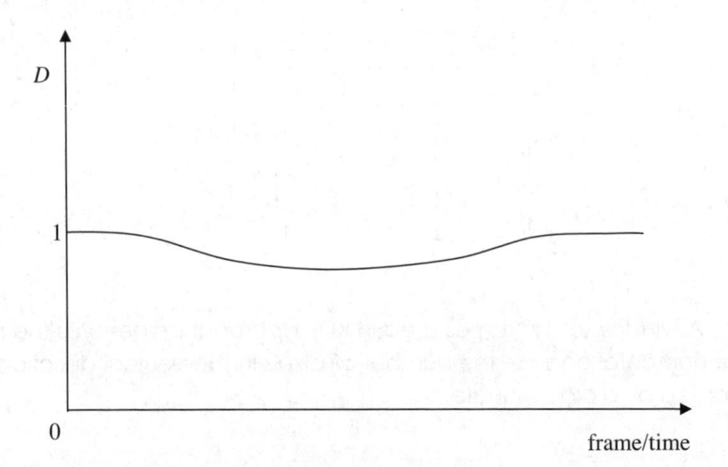

Figure 13.3 Density drop that represents the change in illumination is not sudden and is global (applies to all pixels)

misinterpreted as a foreground. This challenge can be addressed by using the fact that it is usually not sudden, but rather prolonged and it affects all or most of the pixels and is, thus, global, while a novelty (new object) is usually local in the image frame and affects a relatively small number of pixels.

RDE can be useful to eliminate the effects of change of illumination by proving an adaptive (not static) bg model in terms of density, see Figure 13.3.

The approach to distinguish between a novelty and illumination changes can be formulated as follows: If such a smooth (not sudden and significant) drop is detected in all or most of the pixels of the image frame this represents a change of the illumination of the environment rather than a new object that appears on the scene.

13.2.2.2 Shadows and Reflection

Shadows that are cast by the moving objects usually differ from the background and can be misclassified as a foreground, which would increase the false positives (FP). In addition, reflections from wet surfaces, glasses, windows lead to an even more acute problem, because, unlike shadows, they retain colour, texture and edge information that is required by most algorithms for background subtraction.

One way to address shadows is to use a representation by brightness and chromaticity, HSV instead of RGB. The shadows affect mostly the brightness and not so much the chromaticity (Porikli and Tuzel, 2003). Reflections are more difficult and methods to address it rely on using 3D maps, stereovision or thermal imagery to aid the scene interpretation.

From the point of view of density representation, shadows and reflection will lead to an area of pixels having a drop in the density, but unlike real novelty objects, the drop will be less significant (colour will be dark) movements of these pixels will be significant (or linked) to the movement of the object, which is unlikely for two independent objects.

13.2.2.3 Occlusions and Camouflage

Occlusions occur when two or more objects pass close by or are undistinguishable from the view of the camera. Camouflage is related to the foreground having the same or very similar colour and/or texture as the background, in general.

A technique to tackle occlusion is to use motion information and track objects. Camouflage can be addressed using thermal channels and colour values.

13.2.2.4 Nonstatic Background and Camera Oscillations

In practice, cameras are often shaken by rain, wind or other factors that leads to comparing pixels with same positions, for example $(i,j)^{th}$ in neighbouring frames to actually represent not precisely the same point of the background. In addition, the background itself is not perfectly static – for example, moving clouds in the sky, trees being blow by the wind, a car or large object being permanently moved from the scene, and so on. Moreover, these oscillations or background nonstationarity are often nonpredictable.

An approach to address this problem is by using local models (Christiani *et al.*, 2010). A similar approach to the one presented in Figure 13.3 can be applied to a nonstatic background as well. Then, such a smooth drop of the density will be visible in an area of the image (not all pixels but some of them).

13.2.3 Parametric versus Nonparametric Approaches

Statistical modelling techniques can be broadly divided into:

a. parametric; and
b. nonparametric (or parameter free).

All parametric approaches assume that the data come from (possibly a set of) random distributions such as the normal/Gaussian ones that can be parameterised by two parameters (mean and covariance). In the context of evolving and autonomous systems this translates to learning both the model structure and the parameters from the data. Nonparametric approaches, on the other hand, do not make such assumptions.

In this context the background subtraction method can be considered as a density thresholding approach. In addition, parametric approaches usually assume a stationarity or *ergodic* nature (Duda, Hart and Stork, 2000) of the video stream, which means that the ensemble average is assumed to coincide with the time average. This assumption allows estimation of the density numerically from the available training data sequence rather than from many parallel video processes with the same statistical characteristics that, in practice, will not be exactly the same. Obviously, this is seldom true and one has to take nonstationarity into account.

More advanced versions of the parametric methods include so-called Gaussian mixture model (GMM) approach where a non-Gaussian distribution can be approximated using parametric local submodels. At each step of this approach two subtasks are addressed:

i. the current data sample (features, e.g. colour of the current pixel) to be assigned to the best matching distribution; and
ii. update the submodels parameters.

In the so-called hidden Markov models (HMM) approach the transition probabilities are being modelled between predefined states such as background, foreground, shadow (Rittscher *et al.*, 2000). Learning HMM that includes probabilities of transitions between states is traditionally done by offline methods such as Viterbi algorithm (Rabiner, 1989) or more recent recursive algorithms such as (Rittscher *et al.*, 2000; Filev and Kolmanovsky, 2012). GMM with evolving structure were considered by Ramezani *et al.* (2008) and Sadeghi-Tehran, Angelov and Behera (2011). An interesting option would be to consider evolving HMM where new states are added or states become obsolete. In some respect such an approach is the one presented by Stenger *et al.* (2001).

Parameter-free (or nonparametric) approaches include, in general, so-called particle filters (Arulampalam, Maskell and Gordon, 2002) and kernel density estimation (KDE) techniques. KDE will be considered and used further as already described in Section 2.4 of the book. As opposed to parametric approaches parameter-free techniques do not require *prior* assumption of the distribution of the statistical properties of the data samples. These approaches are very accurate and flexible, but require a lot of memory (sometimes prohibitively much) to store past image frames. This problem is lessened to some extent by using windows of frames of size W, parallelisation of the algorithms using specialised hardware and other simplifications.

13.2.4 Kernel Density Estimation Method

The KDE approach is often used for separating the *background* from the *foreground* in video streams processing (Elgammal *et al.*, 2002; Zhivkovic and Van der Heijden, 2006). A normal Gaussian distribution is usually assumed (determined by the specific kernel used, see also Section 2.4) and applied to the $(i,j)^{th}$ pixel:

$$p(z_{ij}) = \frac{1}{k}\sum_{l=1}^{k} K\left(\frac{z_{ij}^{*} - z_{lij}}{\sigma}\right) \tag{13.1}$$

where

$$z_{ij} = [z_{1ij}, z_{2ij}, \ldots, z_{tij}, \ldots, z_{wij}]^T; \quad z \in R^n;$$

σ is the bandwidth; n denotes the number of features (for example, if we use colour intensity values, R, G, B – red, green, blue or H, S, V – hue, saturation, brightness $n = 3$);

W – denotes the number of consecutive frames of the window that are used (memorised); I denotes the horizontal and j – vertical position.

Pixels with high value (usually above a certain predefined threshold) of the pdf are assumed to be *background* and *vice versa* for the *foreground* (Elgammal *et al.*, 2002):

$$\begin{aligned} &IF\ (p(z_{ijt}) < \varepsilon) \\ &THEN\ (z_{tij}\ is\ fg) \\ &ELSE\ (z_{tij}\ is\ bg) \end{aligned} \tag{13.2}$$

where ε denotes a prespecified threshold value.

This approach is applicable offline only, because it requires a window (usually a few dozen) of frames each containing, usually, a million pixels to be stored and manipulated for a time interval shorter than 40 ms (to get the result before the next image frame arrives, assuming a rate of 25 fps). The computational complexity of KDE can be estimated at $O(nWhv)$, where h denotes the number of pixels in the horizontal and v – in the vertical (Sadeghi-Tehran *et al.*, 2010).

This approach has an additional disadvantage in terms of the user-specific threshold value, ε, that has to be prespecified balancing between sensitivity and robustness. This is in addition to the bandwidth of the kernel function, σ. For example, a narrow bandwidth will increase false positives making the algorithm oversensitive, but too wide a bandwidth would cause increase of false negatives due to smoothing (Sadeghi-Tehran *et al.*, 2010).

13.3 Detecting Visual Novelties with the RDE Method

The RDE method (see Section 2.5) was introduced as a more efficient, recursive alternative to the KDE approach and applied to video analytics by Angelov *et al.* (2008, 2010). RDE reduces radically the computational complexity to $O(nhv)$, that is by W times or two orders of magnitude! For example, for a window with size $W = 60$ the recursive estimation of the density will need 45 times less time and memory! This makes the algorithm applicable for real-time applications and the use of as many previous image frames as needed (with the whole video or, theoretically, with an infinite number of frames).

Once processed, the images are discarded from the memory and only the current image frame is being used plus a small amount of accumulated statistics in terms of mean, μ and the scalar product, Σ, see Equations (2.31) and (2.32).

In addition, using this nonparametric approach removes the need to define user- or problem-specific threshold, which removes subjectivity and reduces the effect of the noise as a factor of detection. For example, if we use a Cauchy kernel there is no bandwidth parameter, σ.

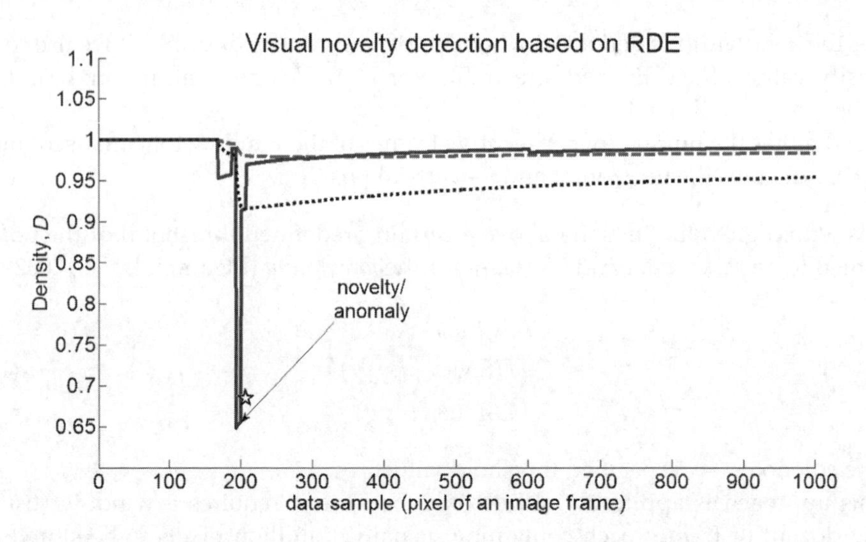

Figure 13.4 Example of the evolution of the density throughout the video-stream for a specific pixel. Frames for which the value of the density drops below the value of $\mu_D - \sigma_D$ are denoted by a star and a novelty is detected there (Ramezani *et al.*, 2008)

RDE applied to video stream works per pixel; for example, for the $(i,j)^{th}$ pixel it can be represented by the following rule, see Figures 13.2 and 13.4:

$$\begin{aligned} &IF\ \left(D(z_{tij}) < \bar{D}(z_{lij}) - \sigma(D(z_{lij}))\right) \\ &THEN\ (z_{lij}\ is\ fg) \\ &ELSE\ (z_{lij}\ is\ bg) \end{aligned} \tag{13.3}$$

where $\sigma(D(z_{lij}))$ is the standard deviation of the densities of image frames seen so far. $\bar{D}$ denotes the mean/average of the global density.

13.4 Object Identification in Image Frames Using RDE

The pixels suspected to be foreground can be identified using RDE. Once this is applied to each pixel of the current frame the next step of the video analysis algorithm (see Figure 13.1) is to identify the physical object(s) that are new to the scene. Traditionally, this is done by clustering or, simply, grouping the suspected pixels often using the mean/average position, see Figure 13.5:

$$\begin{aligned} h_k^* &= \frac{1}{N_F}\sum_{i,j=1}^{N_F} h_{kij} \\ v^* &= \frac{1}{N_F}\sum_{i,j=1}^{N_F} v_{kij} \end{aligned} \tag{13.4}$$

Figure 13.5 On the right half of the figure is the real image and on the left half of the figure are the pixels detected as foreground in black. One can see the pixels on the right top of the left half of the figure are circled with an ellipse that represent noise due to movements of the leaves of the tree. The two squares indicate the centre of the target as identified by the traditional approach (spatial mean of the suspected pixels) and RDE, respectively

where N_F denotes the number of pixels in a frame classified as foreground ($N_F \ll hv$).

The traditional approach is straightforward, but can often lead to problems in identifying the precise position of the object due to the presence of noise that acts as false positives and the lock on the target may be shifted even into some infeasible locations, see Figure 13.5 that is reproduced form (Angelov, Ramezani and Zhou, 2008).

Alternatively, if RDE is applied to the suspected pixels only spatially (in terms of the vertical and horizontal positions within the frame not between frames as it is in Figure 13.2) a much better lock on the target can be achieved as seen in Figure 13.5. The pixel that is closer to most of the other suspected pixels (that is, has higher density) is considered to represent the physical object:

$$T_t^* = \arg\max_{i,j=1}^{N_F}\{D_{tij}\} \\ T_t^* = \left[h_t^*, v_t^*\right] \tag{13.5}$$

where

T_t^* denotes the vector of the object/target position in the current;
t^{th} image frame with its horizontal and vertical components.

The rationale of such an approach is that pixels that are detected due to noise, not because they represent actual physical objects are more likely to be dispersed and not concentrated in a local area, while the physical objects, on the contrary, are grouped

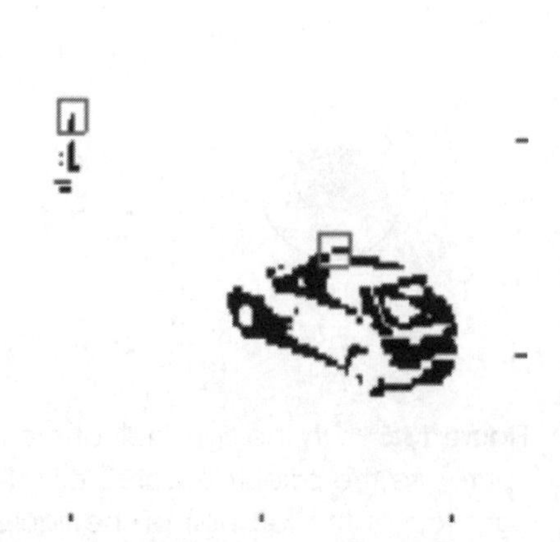

Figure 13.6 Background subtraction using RDE method, Left-hand side plot represents the scene in which there are two moving objects new to the scene; right-hand side plot represents the foreground in black and the background in white. The squares denote the focal points of the physical objects identified to be moving (the car and the pedestrian, respectively)

closely together. This results in a more robust identification of the point/object that is to be tracked. Identifying multiple objects (Figure 13.6) can be done using clustering of the foreground pixels only (Sadeghi-Tehran *et al.*, 2010).

Typically, in video analytics this step follows the step of detecting the foreground pixels, but it can also be used as a standalone approach for image segmentation, landmark detection or SLAM based on visual information.

13.5 Real-Time Tracking in Video Streams Using ALS

As illustrated in Figure 13.1, the first step in video stream processing is detecting the foreground (which was described in Section 13.3) followed by identifying the physical object/target (which was described in Section 13.4) and then upgraded by the higher level of analysis such as motion tracking from frame to frame and/or in the physical space, trajectory generation and/or classification (Sadeghi-Tehran and Angelov, 2012) and/or other higher level analytical tasks (such as intent identification, face expression etc., for example).

The main goal of object tracking in video streams is to predict the position of the object/target within the next, $(t+1)^{th}$ image frame:

$$\hat{T}^*_{t+1} = f(T^*_t) \tag{13.6}$$

where $\hat{T}^*_{t+1}$ is the predicted position of the target/object in the $(t+1)^{th}$ frame.

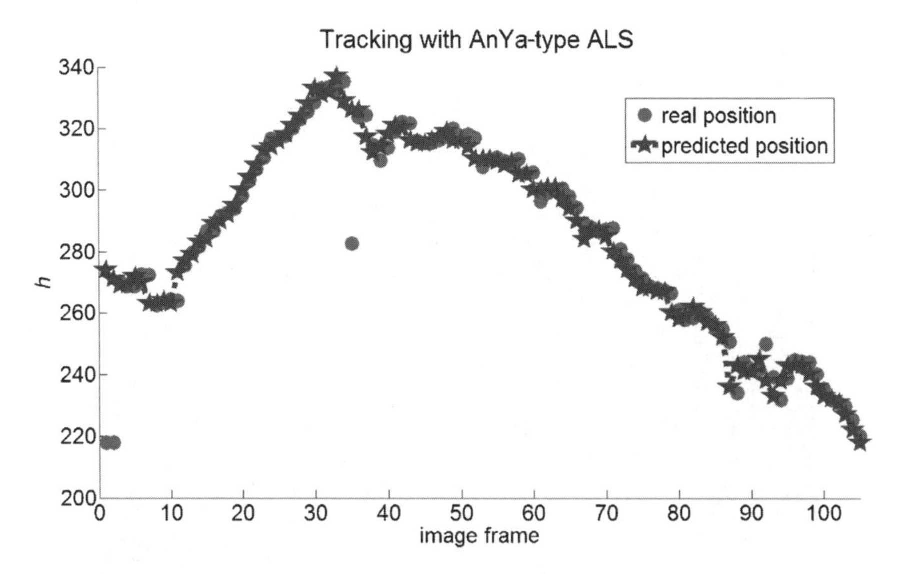

Figure 13.7 Tracking performance compared with the real data for the horizontal component of the motion

In practice, only the centre of the object (as described in Section 13.4) is being tracked. An approach that is widely applied for online tracking is the so-called Kalman filter (Kalman, 1960) that assumes a linear model of the target movements, Gaussian distribution and independence of the parameters and can be used in a Gaussian mixture models (GMM) if we use local submodels and the *divide et impera* principle. If, in addition, we use an FRB system as a framework for ALS one can extract an interesting linguistic interpretation of the prediction models in the following form

$$\begin{aligned} &Rule: \quad IF\left(\begin{pmatrix} h_t^* \\ v_t^* \end{pmatrix} \sim \begin{pmatrix} \chi^* \\ \omega^* \end{pmatrix}\right) \\ &THEN \begin{cases} \hat{h}_{t+1} = a_{10} + a_{11}h_t + a_{12}v_t \\ \hat{v}_{t+1} = a_{20} + a_{21}h_t + a_{22}v_t \end{cases} \end{aligned} \tag{13.7}$$

where

a represents the parameters of the (linear) consequents;
h denotes horizontal and v – vertical position within an image frame.

Several such rules (local submodels) can be extracted and evolved autonomously from the video-stream, see Figure 13.8. Due to the evolving nature of the ALS the number of such rules is not fixed.

Figure 13.8 Focal points (shown with stars) of the local submodels (rules) for the object tracking using AnYa-type model, see expression (13.9)

The overall prediction of the position of the object/target in the next image frame can be produced using a weighted sum following Equation (4.6), Figure 13.7:

$$\hat{T}^*_{t+1} = \sum_{l=1}^{R} \lambda_l T^*_{tl} \tag{13.8}$$

where

λ_l is the normalised firing level of the lth local submodel;
T^*_{tl} is the prediction by the lth local submodel (rule).

The performance of the RDE approach was compared with the traditional BS and KDE approaches and proved to be superior. RDE is more robust, threshold-free, and computationally more efficient. The tests reported by Ramezani *et al.* (2008) demonstrate that RDE is orders of magnitude faster and requires orders of magnitude less memory than the traditional KDE-based approach.

An important advantage of using FRB framework ALS specifically is that the result can be represented linguistically in a transparent and interpretable manner (Angelov,

Ramezani and Zhou, 2008), for example:

$$
\begin{aligned}
&Rule_1 : IF \left(\begin{pmatrix} h_t^* \\ v_t^* \end{pmatrix} \sim \begin{pmatrix} 283 \\ 9 \end{pmatrix}\right) \; THEN \begin{cases} \hat{h}_{t+1} = -4.77 + h_t + 0.068v_t \\ \hat{v}_{t+1} = 20.75 - 0.03h_t + 0.84v_t \end{cases} \\
&Rule_2 : IF \left(\begin{pmatrix} h_t^* \\ v_t^* \end{pmatrix} \sim \begin{pmatrix} 301 \\ 135 \end{pmatrix}\right) \; THEN \begin{cases} \hat{h}_{t+1} = 5.04 + 0.98h_t + 0.01v_t \\ \hat{v}_{t+1} = 31.48 - 0.10h_t + v_t \end{cases} \\
&Rule_3 : IF \left(\begin{pmatrix} h_t^* \\ v_t^* \end{pmatrix} \sim \begin{pmatrix} 333 \\ 354 \end{pmatrix}\right) \; THEN \begin{cases} \hat{h}_{t+1} = 50.37 + 0.8h_t + 0.05v_t \\ \hat{v}_{t+1} = 248.54 - 0.94h_t + 1.18v_t \end{cases}
\end{aligned}
\qquad (13.9)
$$

These are three fuzzily blended KFs that are responsible for their respective local area defined by the antecedents as vicinities of the respective focal points. Jointly, however, they cover the entire image and, especially, the areas where the physical object passes by, see Figure 13.8.

13.6 Conclusions

In this chapter applications of RDE for foreground detection in a computationally more efficient way ($3W/4$ times faster and using $3W/4$ less memory than the traditional BS and KDE approaches) were described and illustrated for the example of video analytics.

The use of RDE in terms of spatial density for a more precise identification of the focal point of physical objects to be tracked was also presented in Section 13.4. Finally, an alternative method for tracking in video-streams based on ALS of AnYa type was presented that has a linguistic and multimodal (GMM-like) nature, but is extracted (and evolved) autonomously from the video stream.

The proposed ALS of AnYa type with the use of RDE offers the opportunity to perform video analytics in real time without direct human involvement, faster, using less memory, no user- or problem-specific thresholds. It is, thus, instrumental for a range of security and surveillance as well as mobile robotics applications.

14

Modelling Evolving User Behaviour with ALS

Human (including user) behaviour modelling is a challenging task and a complex phenomenon that is very important nowadays from many perspectives:

- quality of services provided;
- available technology (ubiquitous sensors, computational and wireless communication devices) makes this dream a reality;
- privacy and security;
- physics, physiology and psychology, medicine and care;
- mathematical modelling, learning, and so on.

In this chapter only the last aspect will be the focus of consideration. The ALS proposed in the book will be studied as a useful tool for modelling user behaviour in its dynamic evolution. This problem is important for computer users, users of the Internet, for the so-called assisted living directed towards the increasingly older population in the developed world and in the context of the ubiquitous computing, communication and sensory devices that are becoming an everyday reality.

14.1 User Behaviour as an Evolving Phenomenon

Traditionally, user behaviour is considered as a complex, but **fixed** category. The important aspect of its dynamically evolving nature is often completely ignored. In reality, the behaviour of users is a dynamic category that may change with time, it can be erratic or influenced by other factors (Iglesias *et al.*, 2012).

A brief review of the existing literature indicates that the methods and techniques proposed and used ignore the evolving nature of the problem. For example, (Han and Veloso, 1999) use Hidden Markov Models (HMM) to recognise behaviours of agents and each state of the HMM corresponds to an abstraction of the behaviour

Autonomous Learning Systems: From Data Streams to Knowledge in Real-time, First Edition. Plamen Angelov.

of the agent. The structure of the HMM is, however, predefined and the training of the HMM is performed offline based on a precollected data stet. In addition, the Markovian assumption (which is necessary in order to facilitate the computational derivation means that the current sate depends directly only on the previous one, which is an idealisation.

The approach used in ALS as presented earlier does take into account the ordered sequence of events. In another study, Godoy and Amandi (2005) present an approach for user profiling based on the use of the Internet, but they consider a batch set of data available offline, which is not realistic in a real-life scenario.

The ability to recognise human behaviour in real time and to classify it without being limited to a pretrained sample data, prespecified number of classes or data patterns is of immense importance for the contemporary intelligent services and devices. Typically, this task is performed by a software agent that gets as inputs the sensory data stream in real time, but is most often pretrained to recognise only a specific set of behaviours or actions (Pepyne, Hu and Gong, 2004).

In the so-called ambient assisted living scenario, but also in general, accurate user behaviour modelling can be useful for predicting the reactions of users to a certain environment or offering. So-called user *profiles* are traditionally hand crafted and rely heavily on expert knowledge (Pepyne, Hu and Gong, 2004). It can be defined as a collection of user interests, preferences, characteristics, behaviours (Iglesias *et al.*, 2012).

In fact, behaviour is only one of the facets that is manifested by the user activity and in a nowadays digital world it can relatively easily be detected, parameterised, recorded or processed digitally. The problems are related to the fact that there are more subtle elements that describe the user such as goals, intention, sometimes masquerading or mimicking someone else, emotions, and so on.

Therefore, the ALS proposed in this book are particularly suitable for the uncertain and dynamically evolving nature of the problem, which boils down to extracting knowledge from an evolving data stream and in particular to classifying it into a not necessarily predefined number of classes. The proposed solution should address the following characteristics:

- should be able to learn additional information from new data;
- should be able to identify and accommodate if a new class is needed based on the data pattern;
- should preserve the previously acquired knowledge in a human intelligible form;
- should not require access to the past historical data (to avoid overload they will not be stored in the memory).

14.2 Designing the User Behaviour Profile

As a first step towards the online design of a user behaviour profile from streaming data one needs to convert the raw sensory data into a form suitable for the classifier. This is usually a stage of preprocessing that is traditionally done offline based on a batch set of training data. The disadvantages of such an approach are obvious and, therefore, here an alternative proposed by Iglesias *et al.* (2009) will be briefly described.

The data stream that describes user behaviour can be considered as a sequence of ordered events. For example, they can be labelled by letters and then the data stream will look like a natural text with words and sentences. In different applications these events take different forms.

For example, if one describes a user of a computer with UNIX operating system, the events may represent, simply, typing different UNIX commands in the command line (similarly, in Web browsing). In another scenario of a *smart* home (Badami and Chbat, 1998) these may represent activities of a daily life, such as making a phone call, washing hands, cooking, eating, and so on. Equally, this can be represented by a sequence of sensor readings which can be binary (ON/OFF) (Iglesias *et al.*, 2010) or continuous (Andreu *et al.*, 2011).

Iglesias *et al.* (2009) proposed to use a *trie* structure to represent the sequence of events in the user's contact with the computer or *smart* home or another *smart* device. This structure itself is not new – it was described first by Fredkin (1960), but its attractiveness for such problems, in particular, is linked to the fact that it can easily be updated recursively (thus, it is computationally very efficient). It also represents quite well the human behaviour in its variety and keeps the order-dependency statistics. As a second stage, statistical characteristics of the *trie* structure are extracted and updated also in real time and recursively.

Finally, the ALS, namely *AutoClassify*, as described in Chapter 8, is applied and linguistic, easy to interpret rules are derived (extracted) from the data stream and evolved. Class labels are assigned automatically and can be confirmed or amended by a human operator whose role is only to monitor the process.

Let us consider a user of a computer with UNIX operating system and the sequence of commands that this user types in. This may be, for example:

$$\{date \rightarrow ls \rightarrow mv \rightarrow ls\} \tag{14.1}$$

The first step is to segment the sequence of events, E into subsequences of equal length, l. For example, if we have the sequence:

$$E = E_1 \; E_2 \; \ldots \; E_n \tag{14.2}$$

then, we aim to get subsequences of the following form:

$$E_i \; \ldots \; E_{i+l} \tag{14.3}$$

In the case study described later, the length of the subsequence considered is $l = 3$ or $l = 6$, but this is very much problem dependent. For the example provided in (14.1) for $l = 3$ one will have:

$$\{date \rightarrow ls \rightarrow mv\} \tag{14.4}$$

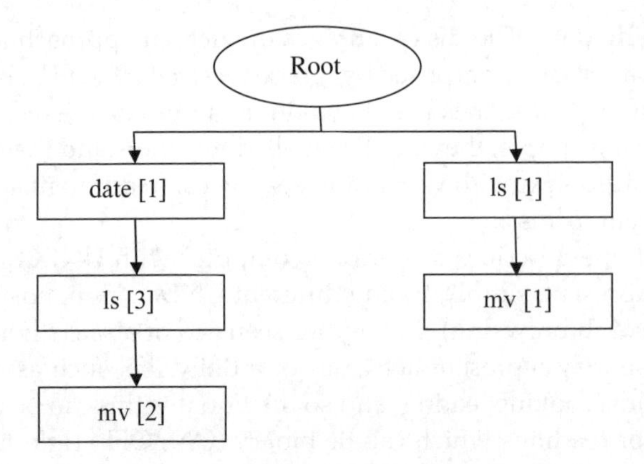

Figure 14.1 An example of a *trie* structure

and

$$\{ls \rightarrow mv \rightarrow ls\} \tag{14.5}$$

In the memory are stored the *trie* structures, not the raw data with the UNIX commands (events), which are significantly more. The *trie* starts at some point to only update the frequencies of appearance of certain combinations rather than to store all of them. However, it preserves the sequential nature of the events that took place (Iglesias *et al.*, 2009).

Every *trie* has a root and children nodes. The first node following the root represents the closing event (the one that is at the end of the sequence). Every children node represents an event that preceded this event and so on (see Figure 14.1 for an illustration of the example shown above).

It is important to note that each node keeps (and updates in real time) the number of times a specific event (in this particular case, UNIX commands typed in the command line by the user; alternatively, this can be the web sites visited, the activities undertaken in a *smart* home environment, etc.). Moreover, this *trie* structure is evolving with time to reflect the events that take place and, in particular, if a new event takes place that does not exist in the trie structure a new root and a node is being added and a new *trie* structure starts to grow/evolve.

The *trie* structure updates the statistics/relative frequency of appearance of each event, but, more importantly, it also takes into account the sequence under which these events take place, which is not the case with the 'simple'/traditional statistics.

Once the *trie* is available the user profile is designed based on the relative frequency of the appearance, see Figure 14.2. The relative frequency is defined by Iglesias *et al.* (2010) as a ratio of the number a particular subsequence of events towards the total number of subsequences of the same length.

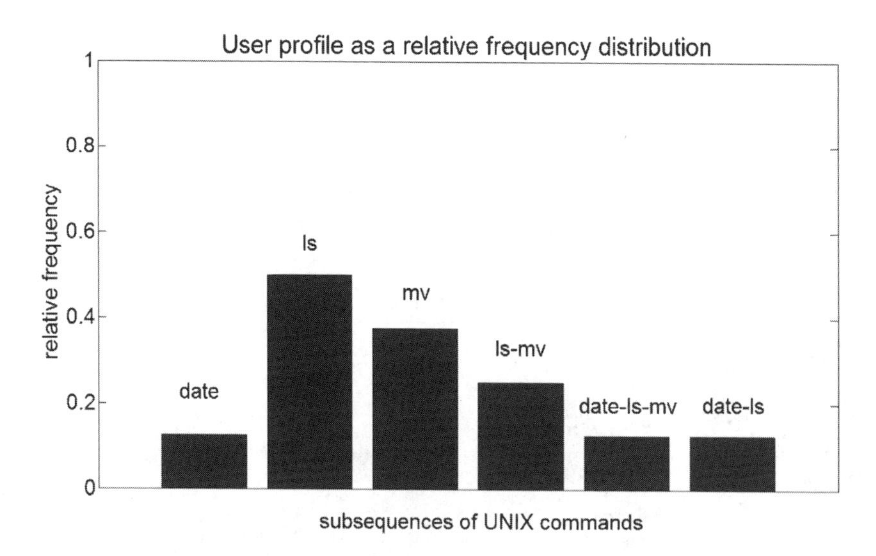

Figure 14.2 Relative frequency used as a basis for designing user profiles

In this way, the *trie* structure can be represented by a distribution of the relative frequencies. Iglesias *et al.* (2012) propose to consider the user profile as a matrix composed of the distributions of relative frequencies as described above, see Figure 14.3.

14.3 Applying *AutoClassify0* for Modelling Evolving User Behaviour

ALS, and, in particular, *AutoClassify0* are particularly suitable for the task of joint learning and classifying user behaviour profiles into classes that are ***not*** predefined and can expand/evolve. *AutoClassify0* is a 0-order evolving classifier that is completely unsupervised and can expand its rule base and can add labels to the classes *post factum* (*a posteriori* to the observations). The work of *AutoClassify0* is described in Chapter 8 and also in Appendix B6 and illustrated for the case of UNIX commands users in Figure 14.4 in the next section.

It takes as input the user profiles that are derived automatically and recursively updated as described in the previous section. These are composed of relative frequencies of the events distributions. Based on these, in the data space composed of all the n existing subsequences the density of each particular observation is calculated recursively and based on the principles described in Section 5.2.5 data clouds (or clusters) are being evolved.

Each prototype selected based on the procedure described in Chapter 8 and in Appendix B6 initiates a (fuzzy) rule. The consequents of these rules initially are filled in automatically with 'Class i' until a human operator or an automatic algorithm labels them with more appropriate/meaningful labels such as '*novice programmers*',

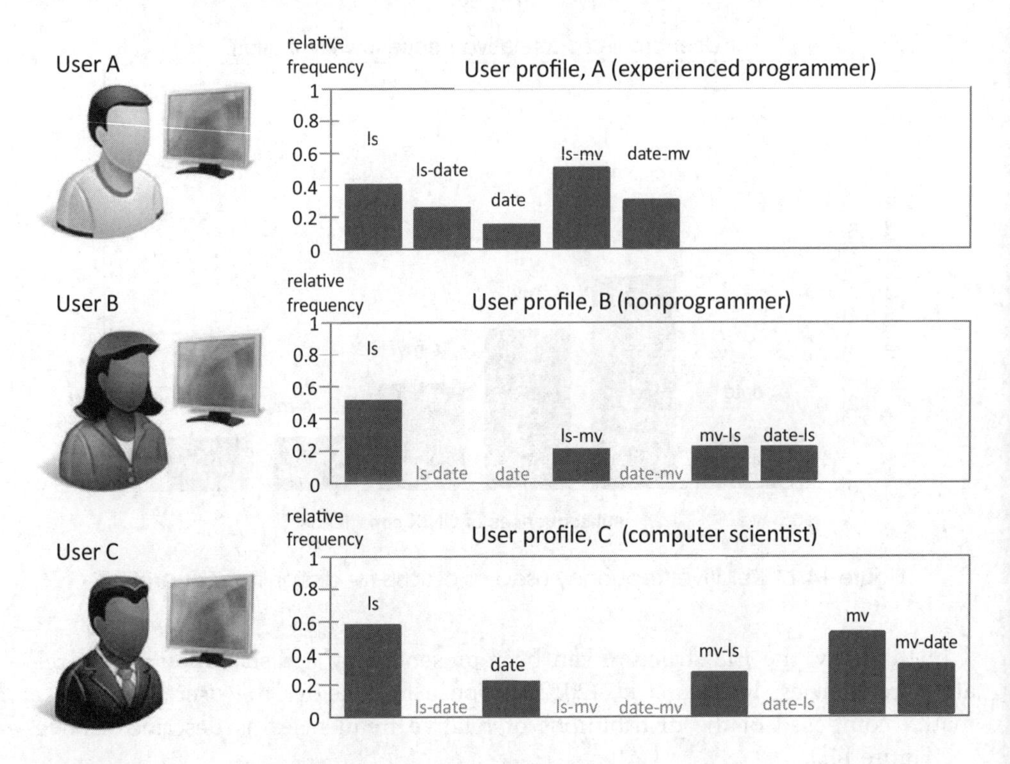

Figure 14.3 User profiles in a matrix form

'experienced programmers', *'computer scientists'*, and so on. It must be stressed that this labelling is not a form of supervision and merely aims to increase the human interpretation capacity of the ALS used, because these labels are not used in the optimisation or other learning scheme.

14.4 Case Studies

14.4.1 Users of UNIX Commands

In this case study a dataset collected form 168 real users collected by Greenberg (1988) was used. They represent several types of programmers, namely

- novice programmers (users with little or no previous experience with programming, operating systems, and UNIX in particular that has specific command line interface);
- experienced programmers (senior graduates from the Computer Science Department who have a significant amount of knowledge and experience with programming, operating systems, and, in particular, with the UNIX environment; these

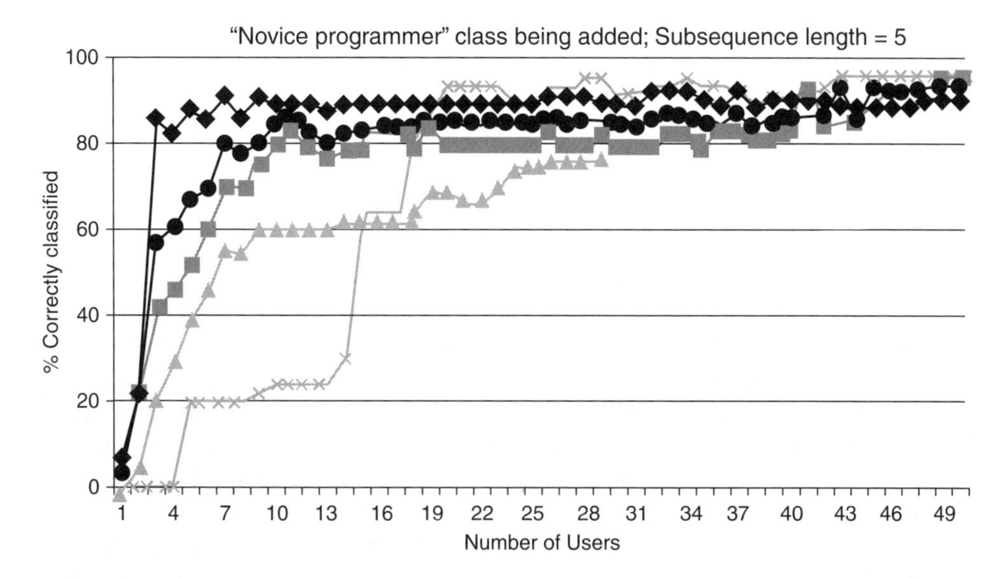

Figure 14.4 Performance of *AutoClassify0* (in line with diamonds) versus the known classifiers (kNN, naïve Bayesian classifier, C5.0) in the scenario when a new class of programmers is being introduced – the curves show the percentage of users of the 'new class' correctly classified (vertical axis) versus the number of users of the new class that contains the training dataset

users have prior experience with coding, word processing, using more advanced UNIX facilities);

- computer scientists (graduates and researchers from the Computer Science Department who have varying experience with UNIX although they are all experienced users of computers, in general);
- nonprogrammers (office staff and faculty members who mostly used computers for word processing and communications and the knowledge of UNIX of whom was the bare minimum needed to do the job).

The available data is summarised in Table 14.1.

Table 14.1 Dataset for the case study of UNIX users behaviour modelling

Users type	# of users	Total # of command lines
Novice programmers	55	77 423
Experienced programmers	36	74 906
Computer scientists	52	125 691
Nonprogrammers	25	25 608
Total	168	303 628

Table 14.2 Results at the end of the run (in %) of applying *AutoClassify0* and other classifiers for subsequence length 6 and 1000 commands used per user

AutoClassify0	C5.0	Naïve Bayes	kNN ($k = 1$)
72.0	74.6	76.1	44.6

It must be stressed that, although, the labels of the groups of users were known they were not used by *AutoClassify0* and only provided for comparison.

The performance of *AutoClassify0* that automatically generated the clusters and (fuzzy) rules was compared with the performance of other well-known and widely used classifiers such as the very powerful tree-based classifier C5.0 (Quinlan, 2003) that is offline and requires all the information including class labels to be known and provided to the classifier beforehand, *k* nearest neighbours (kNN), naïve Bayes, and so on, see Table 14.2.

The following experimental study (Iglesias *et al.*, 2012) includes using the data form one particular class (new to the rule base used so far) to be added to a classifier that has been trained on the data form the other three classes only plus one, then two, and so on samples of the new class. The validation data consists of data form all four classes. The performance of the classifier has been recorded and is depicted in Figure 14.4.

From Figure 14.4 one can see how fast *AutoClassify0* adapts to the new data pattern by evolving its rule base. For the example of *'novice programmers'* after only three new data samples (user profiles) being added in the training phase the ALS is able to predict with nearly 90% precision the users type! Other classifiers not only require a complete retraining but are able to 'catch up' only after seeing a significantly larger number of users (data samples) being provided.

14.4.2 Modelling Activity of People in a Smart Home Environment

The *smart* home scenario is a quite fashionable recent environment for more futuristic research linked to ubiquitous computing, wireless communications and sensory devices. These devices are becoming ever smaller and yet more capable with every year passed.

In this example a dataset collected and reported by CASAS *smart* home project by Washington State University was used (CASAS, 2010). The data concerns sensor measurements from a *smart* apartment – a specially designed facility for experiments in intelligent living environments in Washington State University, USA. These include motion sensors, temperature, water, burner, telephone usage. The data represent 24 users (inhabitants) who perform the following five activities of a daily life:

- making a telephone call;
- washing hands;
- cooking;

Table 14.3 An extract of the sensory records for the daily activity *Cooking* (adopted from Iglesias *et al.* (2010))

Date and time	Sensor readings	Sequence of activities
29/02/2008 13:25:05	101 ABSENT	*I01 ABSENT*
29/02/2008 13:25:09	M16 OFF	*M16 OFF*
29/02/2008 13:25:10	M17 ON	*M17 ON*
29/02/2008 13:25:11	I07 ABSENT	*I07 ABSENT*
...	...	...

- eating;
- cleaning.

The overall dataset consist, thus, of 120 different subsets labelled by one of the above labels where each subset consists of 30 to 150 sensor readings. For example, in Table 14.3 an extract is shown of the daily activity *Cooking*.

In this table M16 and M17 represent motion sensors and I01 and I07 represent a sensors that are activated when a person or an item is present in the room and the default alternative value is ABSENT.

In a similar manner to the user behaviour analysis of the UNIX commands the ALS of type *AutoClassify0* was used. The results were, again, compared with the results of existing state-of-the-art classifiers including offline high performers such as C5.0 that require all the training data to be available including class labels and perform many iterations as well as incremental classifiers such as *k*NN and naïve Bayes that can work on a sample by sample basis, but do not evolve and the number of classes is also required to be known and provided to the classifier beforehand. The results are tabulated in Table 14.4.

The results demonstrate that a comparable and superior precision can be achieved, but with no prior knowledge of the class number or labels and the ability to add new classes as they arrive.

14.4.3 Automatic Scene Recognition

Finally, another interesting application is represented by an automatic system for learning scenes from images. Nowadays, the spread and importance of social web sites and media such as Facebook and the like is growing fast and the amount of images loaded on the Internet is exponentially growing as well. Processing images

Table 14.4 Results – classification rate (in %) of applying *AutoClassify0* and other classifiers for subsequence length 3

AutoClassify0	C5.0	Naïve Bayes	kNN ($k = 1$)
94.2	95.0	88.3	86.6

is, in principle, a computationally expensive task that is rarely performed in real time and, especially, recognition and association of the images with elements of the environment or objects associated by the humans with some specific type of environment.

In this line of thinking, the application to automatically learn and recognise images is very attractive. The proposed ALS is able to automatically recognise if a specific image is taken, for example, on the beach, in a rural environment, in a forest, in a city or in a small village. That is very attractive for numerous applications, including, but not limited to miniaturised specialised devices, online service accessible remotely or to a standalone software or hardware implementation.

In the research literature there were attempts to address this problem by using graphical models (Murphy, Torralba and Freeman, 2004). Alternatively, Ankenbrandt, Buckles and Petrya (1990) tried to split the scene into semantic concept nets. This required dividing the image into grids of subregions linked with the semantic nets. Oliva and Toralba (2001, 2002, 2006) applied the so-called GIST approach for feature extraction (Li *et al.*, 2008) based on Gabor spectral representation (Gabor, 1946).

It may be strange that the humans are able to recognise complex scenes by just glancing on an image. The rational thinking cannot capture and process complex images in one simple glance. But the human brain captures a spatial representation of scenes and recognises meaningful objects and salient features of the scene. It constructs a representative picture that consists of low-level perceptual and conceptual (such as colour, contours), intermediate (such as shapes, texture regions), and high-level (semantic) information (Andreu, Angelov and Dutta Baruah, 2011).

GIST is designed to provide a low-dimensional representation of the scene in terms of universal coordinates without partitioning (Murillo and Kosecka, 2009). The features that are derived represent perceptual dimensions such as roughness, openness, naturalness, expansions, which represent the spatial structure of the scene (Murillo and Kosecka, 2009).

Torralba *et al.* (2000, 2003) studied methods to satisfy the fingerprint of an image in terms of unique features by creating a structured representation of the scene. For example, urban zones are structured vertically and forests are textured. But different from many other approaches that partition the scene into bins, GIST considers the whole scene and assigns to it a vector of features called a spatial envelope based on spectral representation and coarsely localised information (Murillo and Kosecka, 2009).

The ALS of *AutoClassify1* type with recursive (online) feature extraction (rPCA-rLDA) was applied to a dataset (Oliva and Toralba, 2001), which contains eight types of outdoor scenes (*streets, open country, cities, forests, coasts, mountains, high-rise buildings*, and *highways*). In total, the data consists of 2600 colour images with resolution 256 × 256 pixels. The confusion matrix is depicted in Figure 14.5.

The GIST approach produces 513 initial features that were further reduced online to 22 by rPCA-rLDA approach. The classification rates of 70–73% were achieved on average although some types of scenes where correctly classified in over 90% of the cases, such as *cities, open country, mountains, forests*, Table 14.5. Some types of scenes

Figure 14.5 Confusion matrix for the automatic scene recognition using *AutoClassify* type ALS as described above (reproduced from Andreu, Angelov and Dutta Baruah (2011))

Table 14.5 Classification results for the automatic scene recognition using *AutoClassify* type ALS (Andreu, Angelov and Dutta Baruah, 2011)

Method	Classification rate, %	Computational Time, s
AutoClassify1	79.1578	0.00065
kNN	74.3187	0.00031
SVM	79.6646	0.01
Naïve Bayes	80.2935	0.002

are easy to misclassify and difficult to distinguish, such as *coasts* and *highways* that on sunny days may resemble sea water (Andreu, Angelov and Dutta Baruah, 2011).

It should be stressed again that *AutoClassify1* unlike the other methods works online on a per data sample basis evolving the classification rate, while the other methods are offline.

14.5 Conclusions

In this chapter the application of ALS to modelling evolving user behaviour is demonstrated on several case studies. The ability of ALS and *AutoClassify* in particular, to adapt its structure and evolve is particularly suitable for applying it to such complex and dynamic tasks.

In the first case study, real data from 168 users of computers with UNIX operating system were studied. They were from four different groups (*novice programmers*, *experienced programmers*, *computer scientists* and *nonprogrammers*) that were not provided to *AutoClassify0* but were correctly classified in nearly 90% of the cases even if using a very limited number of a certain group (for example, *novice programmers*). ALS of *AutoClassify0* type demonstrates that it can evolve and learn from a very limited amount of data adapting to the changing data density pattern.

In the second case study, again *AutoClassify0* was used for automatically classifying everyday activities performed by inhabitants of a *smart* apartment in the framework of the project CASAS. These activities include but are not limited to phone call making, cooking, eating, and so on. Again, *AutoClassify0* was able to evolve its structure (human intelligible rule base and classification surface) completely autonomously and achieve over 94% correct classification. Alternative classifiers achieved either marginally higher or lower classification rate, but, more importantly, they required the number of classes and their labels to be provided and for the C5.0 classifier (which is the only one to marginally outperform *AutoClassify0*) a batch set of training data has to be provided as well.

Finally, *AutoClassify1* type ALS is applied for a very interesting problem of automatically recognising the type of a scene from an image. Again, the autonomous classifier has very high performance (on average over 70% correct classification with some types of scenes being recognised with over 90% correct classification).

This again was compared with the most competitive alternatives that again require the number of classes to be prespecified, and in the case of SVM, the batch set of training data. The computational burden of the proposed *AutoClassify* type ALS (measured here by the computational time spent) is significantly lower due to the recursive calculations performed, which makes it a viable option for implementation in a range of real-life devices and systems.

15

Epilogue

This book was planned some five years ago as a vehicle to put in one place the theory and applications of the innovative research, but it evolved in parallel with the development and maturing of the investigations that led to some new and interesting results. The book is structured in three parts; Part I – Foundations, Part II – Methodology of ALS, and Part III – Applications of ALS.

Part I itself is composed of three chapters that represent the foundations of the main cognate areas of research, including probability theory, machine learning and pattern recognition, and fuzzy sets theory (including NFS). Although, this part of the book is rather introductory, new and original ideas and elements are presented, including the powerful (patent pending) concept of RDE, a new method for evolving clustering, ELM and the recent ground-breaking method for fuzzy and neurofuzzy systems modelling, AnYa.

Part II contains the theoretical basis of the innovative and powerful approach of ALS. In particular, Chapters 5 and 6 describe the principles and methodology for autonomous learning of system structure and parameters from data streams. Chapters 7–9 describe predictors, estimators, filters, autonomous learning sensors, classifiers and controllers using ALS. Chapter 10 describes the principles and procedures for collaborative ALS.

Finally, Part III consists of four chapters that describe various applications of ALS to areas as diverse as chemical and petrochemical industry, mobile robotics, autonomous video analytics and user-behaviour modelling.

15.1 Conclusions

This book starts with introducing an innovative and powerful concept – autonomous learning systems. For many years researchers aimed to achieve self-learning and self-organising systems. For example, as far back as 1968 the Russian/Soviet scientist Yakov E. Tsypkin said: *'To solve a real-time control problem it is necessary to determine*

Autonomous Learning Systems: From Data Streams to Knowledge in Real-time, First Edition. Plamen Angelov.

the current characteristics of the object. The controller, as it proceeds with its control actions, is undergoing a self-learning process to learn about its relationships with the objects and about the events that influence a change of these relationships. Over the course of doing that, the controller will discover laws that control events in the best possible way, and be even able to predict further development of events.' Further in the same seminal work, Tsypkin continued: '*We cannot control an object in an optimal fashion without knowing its characteristics, but we can study a plant by controlling it. This can then give us an opportunity to improve the control and eventually to reach an optimal control law, thanks to simultaneous identification and control of the plant. Thus, algorithms of learning and control are closely related to each other that is depend on each other and are inseparable.*'

However, it took decades before the required advances in other areas of research and engineering practice made possible nowadays not only to dream about but to formulate, test, simulate and even run (in the case of mobile robots or motes, smart phones, UAVs, industrial plants equipped with intelligent sensors, etc.) real applications. In the introductory, first chapter the advances and related problems of the cognate disciplines are described, such as machine learning, system engineering (specifically, system identification), data mining, statistical analysis, pattern recognition, including clustering, classification, fuzzy logic and fuzzy systems, including NFS, and so on.

The overall concept of ALS, introduced in this book, can be summarised and visualised as shown in Figure 15.1.

In Chapter 2 the basics of probability theory were critically described in a very brief form that makes possible their later use and comparison with the other techniques that were also considered such as fuzzy systems and neural networks. An important and significant innovation is made by introducing in a systematic way of the

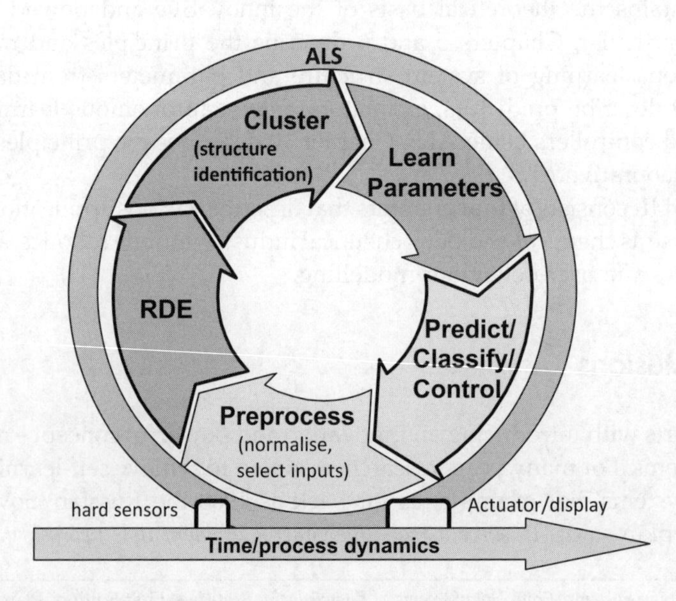

Figure 15.1 The concept of ALS visualised as a never-ending autonomously evolving process

concept of the recursive density estimation (RDE). It has great potential in detection of novelties/anomalies (Kolev *et al.*, 2012a, 2012b) as well as for system structure identification.

In Chapter 3 another well-established discipline, machine learning and pattern recognition was briefly introduced. Again, innovative evolving clustering approach was introduced in which the number of clusters is not prespecified, but dynamically develops as a function of the density in the data space – in particular the ELM approach that is an evolving version of the *mean shift* clustering approach.

The recently pioneered by the author concept of evolving classifiers, which works in a similar manner to the adaptive control systems and estimators by pairs of 'classify and update' actions for each new data sample (or for these new data samples for which class label is known) is also innovative. In this book a particular novel optimal classifier is also introduced based on the quadratic Fisher discriminant analysis which is subject to pending patents.

Chapter 4 provides a concise summary of the theory of fuzzy systems and related NFS. It is, obviously, not possible and it was not the main aim to go into much detail (e.g. operations with fuzzy systems, linguistic hedges, fuzzy equalities and inequalities, fuzzy relational models, etc.) and the interested readers can find these details in other more specialised books. The details that are presented on both FRB and NFS are precisely those that are needed for the further consideration of ALS, which is the topic of the present book.

At the same time, in this chapter, again, novel and recent concepts were introduced, such as AnYa-type FRB/NFS that offer a simpler, yet powerful, description if compared with traditional Mamdani- and TS-type FRB/NFS. Interesting parallels between the Bayesian-type probabilistic models and FRB and NFS of TS and AnYa type were also made.

Chapter 5 starts the second part of the book with the description of the concept of system structure evolution. The key role of the data space partitioning in the process of autonomous system structure design is explained and detailed. Methods for autonomous monitoring the quality of the local submodels measured by their *utility*, *Age*, radii (in the case of clustering) were described as well as methods for real-time, online normalisation and standardisation of the data.

It should be stressed that these principles apply to various type of systems, including, but not limited to fuzzy rule-based, Bayesian, neural network, and so on. The models that are designed in this way are globally nonlinear yet locally can be treated as linear. In addition, ALS of FRB type has an additional advantage – they are also linguistically interpretable, which can be a big advantage for acceptability of these models by human operators.

In this way, a fully unsupervised approach for evolving the system structure autonomously form the data streams emerges that does not require the number of submodels, their focal points or the effective number of inputs used to be predefined – instead it is extracted from the data stream online, in real time.

In Chapter 6 methods and algorithms for autonomous learning of the parameters of evolving systems were introduced. Optimal (both in terms of *local* and *global* criteria) solutions were discussed based on the weighted extensions of the well-known RLS algorithm adapted to the case of multimodel evolving structure. The problem of outliers in learning is also addressed. It has to be noted that while system structure identification is an unsupervised learning problem, parameters of the submodels are determined by semisupervised learning.

In Chapter 7 problems of time-series prediction, filtering and estimation (which are closely related to each other) are described as well as self-calibrating autonomous sensors that play an important role in signal processing, statistical analysis, econometrics and other disciplines and have significant commercial importance, especially in chemical, petrochemical and related process industries.

In Chapter 8 autonomous classifiers are described that form a family *AutoClassify*, including the zero-order *AutoClassify0* and the first-order *AutoClassify1* (both, MISO and MIMO versions). This type of classifier is innovative and pioneering in its own right. Their applications are discussed in more detail in Chapters 13 and 14.

In Chapter 9 the autonomous learning controller, *AutoControl* has been described, which does not require the model of the plant nor previous knowledge about the control policy to be known. The controller structure is self-developed (optionally starting 'from scratch') based on the density and error information from the history of control process collected during its operation and used recursively (without memorising, but still using the history in full and exactly).

In Chapter 10 the powerful and interesting idea of the team of ALSs that can collaborate is briefly described. It has huge unexploited potential in areas such as mobile robotics, wireless sensor networks and UxVs to name just a few. It has been proven precisely that the same result can be achieved by collaborating ALSs as if they all process the whole data when in reality they only have access to and process only their parts of the data plus respective recursively updated statistical variables as detailed in Chapter 10 and in the US patent # 2010-0036780 granted 21 August 2012 (Angelov, 2006).

Part III of the book is devoted to the applications of ALS in the areas of oil refineries, chemical industry, mobile robotics, computer vision, evolving user-behaviour modelling, and so on. These applications are provided primarily to demonstrate the capacity of ALS and are the result of recent research of the author and his students during the last five to six years.

It should be stressed that the applications that are described in Chapters 11–14 are not exhaustive in any way. They do not close the circle of possible applications or even the circle of applications in which the author of the book was involved.

For example, an application domain that was not described in the book, but details on which can be found by the conference and journal papers by the authors and his associates and students, is biomedical problems. For example, the problem of Fourier transformed infrared, FTIR, spectroscopy classification was addressed using ALS in a series of publications (Kelly *et al.*, 2008, 2010; Trevisan *et al.*, 2012a, 2012b). Decision support systems were developed using ALS by McDonald, Xydeas and Angelov (2008). RDE was extensively used for autonomous anomalies detection in flight data analysis (FDA) within the framework of the EU funded project SVETLANA that involves companies SAGEM, France; NLR, The Netherlands, United

Aircraft Corporation, Russia; Concern Avionica, Russia and Lancaster University (http://www.svetlanaproject.eu/).

Readers who are interested in more details are kindly referred to these and similar papers by other authors, for example in the journal Evolving Systems (http://www.springer.com/physics/complexity/journal/12530) and the IEEE annual conferences on Evolving and Adaptive Intelligent Systems (http://www.uc3m.es/portal/page/portal/congresos_jornadas/home_cfp_eais).

15.2 Open Problems

ALS offer very promising new directions of research. They answer many open questions such as the old dream of designing systems that are more intelligent and self-organising even if they are nonlinear, nonstationary and complex. This applies to problems as diverse as clustering, classification, control, filtering, estimation, time-series prediction, intelligent self-calibrating sensors, collaborative systems and a variety of applications.

However, they are not the end of the road, the highest peak one can reach or the 'silver bullet'. They still leave a number of questions unanswered or partially answered. For example, optimality of the proposed schemes, stability and convergence of control and other schemes.

In addition, applicability and possible constraints of ALS is based on a different model framework, for example hidden Markov models, decision trees, and so on. These are problems worth investing more time and research in and investigating. They were not covered at all or not fully in this book and will be of interest for the further development of this emerging and novel branch of research with huge potential to engineering and other applications and are certainly on the horizon of the author's interests.

15.3 Future Directions

As usual, the time and other constraints did not allow expanding further and this book, although being a complete and well-balanced one, represents only a snap-shot of what is an evolving process of the development of this research area. The next steps (at least form the point of view of the research and study) can be summarised as:

- The optimality of the process is difficult to guarantee globally and overall; The RLS procedure itself is optimal, but once combined with the local models one can have either a locally or globally optimal solution and when the structure starts to evolve/change this is, strictly speaking, not true. Ideally, clustering or data space partitioning into data *clouds* should also be optimal. This is a very challenging problem when the data are streaming and, possibly, nonstationary. The current methodology provides a solution that is optimal subject to the structure resulting from the clustering or data *clouds* and this optimality is partial – whenever a system structure evolution takes place the problem needs to be reinitialised with new optimality conditions. Addressing all these problems is one possible direction for future research.

- Stability is an important problem in control. *AutoControl* is a very attractive and innovative scheme that does not require plant model or control law to be known or predetermined, yet human-intelligible and interpretable control rules can be extracted during the process of control of an unknown plant. However, proof of stability of the algorithm and constraints subject to which this will be valid is left for further research.
- Applying the concept of ALS to other frameworks (different form FRB, NFS) such as hidden Markov models, decision trees, and so on. is also another very promising direction for future research.

Appendices

Appendix A

Mathematical Foundations

A.1 Probability Distributions

- Gaussian (normal)

  The *Gaussian*, also known as *normal* (when the mean is zero) distribution is the most widely used one in practice because of its properties. For example, the so called central limit theorem states that the averages of random variables and signals tend to Gaussian; a sum of two Gaussians is a Gaussian again; the distribution that maximises the entropy for a given variance is the Gaussian; it is robust to linear transformations, and so on.

  For a single variable, x it is defined by two parameters, the mean, μ and the variance, σ^2 or equivalently by its square root, the standard deviation, σ:

$$N(x \mid \mu, \sigma^2) = \frac{1}{\sqrt{2\pi\sigma^2}} e^{-\frac{(x-\mu)^2}{2\sigma^2}} \tag{A.1}$$

The entropy is given by:

$$H[x] = \frac{1}{2}\ln(\sigma^2) + \frac{1}{2}(1 + \ln(2\pi)) \tag{A.2}$$

The inverse of the variance, σ^{-2} is called the precision.

In a vector form, the Gaussian distribution is defined by the n-dimensional vector, μ and an $(n \times n)$-dimensional covariance matrix, Σ as described in Chapter 3. The covariance matrix is by definition symmetric and positive (since it is formed by squares of distances) quantities/elements.

$$N(x \mid \mu, \Sigma) = \frac{1}{(2\pi)^{n/2} |\Sigma|^{1/2}} e^{-\frac{(x-\mu)^T \Sigma^{-1} (x-\mu)}{2}} \tag{A.3}$$

Autonomous Learning Systems: From Data Streams to Knowledge in Real-time, First Edition. Plamen Angelov.

The entropy is then given by:

$$H[x] = \frac{1}{2}\ln|\Sigma| + \frac{n}{2}(1 + \ln(2\pi)) \tag{A.4}$$

The precision is again defined as an inverse, but in this case of the covariance matrix, Σ^{-1}.

- Cauchy
  The Cauchy type of pdf is named after Augustin Cauchy (1789–1857) and is given by:

$$p_C = \frac{1}{\pi\gamma\left(1 + \left(\frac{x - x_0}{\gamma}\right)^2\right)} \tag{A.5}$$

which for $x_0 = 0$ and $\gamma = 1$ simplifies to what is known as the standard Cauchy function:

$$p_C = \frac{1}{\pi(1 + x^2)} \tag{A.5a}$$

The Cauchy function can be considered as an approximation of the Gaussian in a first-order Taylor series (see Chapter 2). The main advantage of the Cauchy function is that it is nonparametric and also that it can be calculated recursively, as shown by Angelov (2004, 2006).

- Epanechnikov
  This type of distribution is named after V. A. Epanechnikov and was published in 1969. It is defined through a kernel:

$$\hat{p}(x) = \frac{1}{kh^n}\sum_{i=1}^{k} K\left(\frac{x - x_i}{h}\right) \tag{A.6}$$

where

$$K\left(\frac{x - x_i}{h}\right) = \begin{cases} \frac{1}{2}V_n^{-1}(n+2)\left(1 - \left\|\frac{x - x_i}{h}\right\|^2\right); & \left\|\frac{x - x_i}{h}\right\|^2 < 1 \\ 0 & \text{otherwise} \end{cases}$$ is the

Epanechnikov kernel function that is symmetric but not necessarily positive and integrates to one;

$h > 0$ is the radius;

V_n is the volume of the unit n-dimensional sphere.

- Student
  The student or t-distribution, as it is known, was proposed by William Gosset in 1908 who published under a pseudonym. It can be seen as an infinite mixture of Gaussian distributions with the same mean and different variances:

$$t(x \mid \mu, \lambda, p) = \frac{\Gamma\left(\frac{p+1}{2}\right)}{\Gamma(1/2)} \sqrt{\frac{\lambda}{\pi p}} \left(1 + \frac{\lambda(x-\mu)^2}{p}\right)^{-\left(\frac{p+1}{2}\right)} \tag{A.7}$$

  where

  $p > 0$ denotes the degrees of freedom (if $p = 1$ the distribution reduces to Cauchy distribution);
  Γ denotes the so-called gamma function (Bishop, 2009);
  λ is the precision.

  The vector/multivariate t-distribution is given by:

$$t(x \mid \mu, \Lambda, p) = \frac{\Gamma\left(\frac{p+n}{2}\right)}{\Gamma(n/2)} \frac{\Lambda^{1/2}}{(\pi p)^{n/2}} \left(1 + \frac{\Delta^2}{p}\right)^{-\left(\frac{p+n}{2}\right)} \tag{A.8}$$

  where

  the Mahalonobis distance is defined by $\Delta^2 = (x-\mu)^T \Lambda (x-\mu)$;
  the covariance is given by $\frac{p}{p-2}\Lambda^{-1}$.

- Uniform
  The uniform distribution is the simplest one. For a single variable, x it is defined as:

$$U(x \mid a, b) = \frac{1}{b-a} \tag{A.9}$$

  where

  the mean, $\mu = \dfrac{a+b}{2}$;

  variance, $\sigma^2 = \dfrac{(b-a)^2}{12}$

A.2 Basic Matrix Properties

- Scalar (or inner) product
  The scalar (or inner) product is defined over two vectors with the same dimension, n (which, in general may be the same) by:

$$\Sigma = x^T y = y^T x = \sum_{i=1}^{n} x_i y_i \tag{A.10}$$

- Eigenvector and eigenvalue
  Eigenvector and eigenvalues are properties of square matrices that have special roles, in particular in the PCA approach described in Chapter 3. They are defined by:

$$Au_i = \lambda_i u_i;\ i = 1, 2, \ldots, n \tag{A.11}$$

where

A is a $n \times n$ matrix;
u_i is the eigenvector, and
λ_i is the corresponding eigenvalue.

This set of n linear equations has solutions (because the order is n there will be n solutions) given by:

$$|A - \lambda_i I| = 0 \tag{A.12}$$

which is also called the characteristic equation.

- Matrix determinant
  The matrix determinant is defined over a square matrix. For example, for a *2D* matrix ($n = 2$) it is defined as follows:

$$|A| = \begin{vmatrix} a_{11} & a_{12} \\ a_{21} & a_{22} \end{vmatrix} = a_{11}a_{22} - a_{12}a_{21} \tag{A.13}$$

The determinant of the inverse matrix is given by:

$$\left|A^{-1}\right| = |A|^{-1} = \frac{1}{|A|} \tag{A.14}$$

- Woodbury lemma concerning matrix inverse
  The so called Woodbury lemma (named after Max A Woodbury) provides the identity for the inverse of a correction to a matrix by correcting the inverse of the original matrix (assuming that the correction is with rank q).

$$(A + BCD)^{-1} = A^{-1} - A^{-1}B(C^{-1} + DA^{-1}B)^{-1}DA^{-1} \tag{A.15}$$

where

A, is a $n \times n$ matrix;
U is a $n \times q$ matrix;
C is a $q \times q$ matrix, and
D is a $q \times$ n matrix.

The Woodbury lemma can be proven and is used in derivation of the results for the RLS (see Chapter 6) and Kalman filter.

Appendix B

Pseudocode of the Basic Algorithms

B.1 Mean Shift with Epanechnikov Kernel

Algorithm 1 Pseudocode of the mean shift algorithm with Epanechnikov kernel

Begin (MeanShift_E)

DO for i=1,...,N

For each data point, $\mathbf{x}_i$, $x \in R^n$ calculate the gradient by $\hat{\nabla} p(x) \equiv \nabla \hat{p}(x) = \frac{1}{Nh^n} \sum_{i=1}^{N} \nabla K\left(\frac{x-x_i}{h}\right)$;

If use Epanechnikov kernel, $K(u) = \begin{cases} \frac{1}{2} V_n^{-1}(n+2)(1-||u||^2), \; if \; ||u||^2 < 1 \\ 0, \; otherwise \end{cases}$

its gradient is given by:

$$\nabla K\left(\frac{x-x_i}{h}\right) = \nabla\left[\frac{1}{2V_n}(n+2)\left(1-\left\|\frac{x-x_i}{h}\right\|^2\right)\right] = \frac{1}{2V_n}\frac{(n+2)}{h^2}\left[-2(x-x_i)\right]$$

The gradient of the pdf then becomes:

$$\hat{\nabla} p(x) = \frac{1}{N2(h^n V_n)}\frac{(n+2)}{h^2}\sum_{i=1}^{N}\left[-2(x-x_i)\right] = \frac{N_x}{N(h^n V_n)}\frac{(n+2)}{h^2}\left(\frac{1}{N_x}\sum_{x_i \in S_h(x)}(x_i - x)\right)$$

where the region $S_r(\mathbf{x})$ is a hypersphere of radius h having the volume $h^n V_n$, centred at $\mathbf{x}$, and containing N_x data points.

The so-called *sample mean shift* ($M(\mathbf{x})$) is then:

$$M(x) = \frac{1}{N_x}\sum_{x_i \in S_h(x)}(x_i - x) = \mu - x$$

Autonomous Learning Systems: From Data Streams to Knowledge in Real-time, First Edition. Plamen Angelov.

where μ is the local-mean i.e. the mean of samples in the region $S_h(\mathbf{x})$.
Taking into account also that the constant term $\frac{N_x}{N(h^n V_n)}$ is the pdf estimate over the region $S_h(\mathbf{x})$, thus $\hat{p}(x) = \frac{N_x}{N(h^n V_n)}$ $\nabla\hat{p}(x) = \hat{p}(x)\frac{(n+2)}{h^2}M(x)$
Each sample is moved towards the mean using gradient ascent with normalized gradient:

$$\mathbf{x}_j^{k+1} = \mathbf{x}_j^k + c\frac{\nabla\hat{p}(x)}{\hat{p}(x)}$$

where $\mathbf{x}_j^k$ = j^{th} data sample at the kth iteration.
Substituting the constant $c = \frac{h^2}{n+2}$, one finally gets (each sample is shifted with a value equal to the local mean):

$$x_j^{k+1} = x_j^k + M\left(x_j^k\right) = x_j^k + \mu - x_j^k = \mu$$

End (MeanShift_E)

B.2 *AutoCluster*

Algorithm 2 Pseudo-code of the *AutoCluster* algorithm

Begin (AutoCluster)
Initialise possibly 'from scratch'; **IF** 'from scratch' **THEN**
 the first data point becomes a focal point, $x_1^* \leftarrow x_1$;
 its density increment is set to 1, $\delta_1 \leftarrow 1$;
 the mean value with the first data point, $\overline{x}_1 \leftarrow x_{11}^*$; $k \leftarrow 1$.
Starting from the next data point ($k \leftarrow k+1$, x_k)
DO WHILE there is no more available data or
UNTIL a requirement to stop the process is received:

1) calculate the density *increment*, δ_k in respect to ***all*** previously existing focal points;
2) update the global mean value (of ***all*** data points), $\overline{x_k}$;
3) **IF** *Principle A (see Section 5.2.6)* holds
 THEN add the new data point as a new focal point, $x^{(R+1)*} \leftarrow x_k$;
 update quality parameters (*Age*, support, utility, radius);
4) **IF** *Principle B (see Section 5.2.6)* holds
 THEN remove the(se) previous focal point(s) that are close to the new data point and replace it/them with the new data point, $x^{i*} \leftarrow x_k$;
 ELSE ignore (do **not** add new focal point) **but** update the quality parameters (age, support, utility, radius);

```
5) IF there is (a) focal point(s) with low utility, support or
   high age
   THEN remove it/them;

End (AutoCluster)
```

B.3 *ELM*

Algorithm 3 Pseudocode of the ELM algorithm

Define $\mathbf{x}_i$ = current data sample, i indicates the instant at which $\mathbf{x}$ has arrived or simply the position in a data sequence, $\mathbf{x} \in R^n$. μ_i = i^{th} cluster centre (local mean). r = radius. σ_i=average from i^{th} centre to all the samples in cluster i. c = number of clusters, $count_i$ = number samples belonging to i^{th} cluster. $\|\mathbf{x} - \mathbf{y}\|$ = norm of vector **x-y**. α_i = sum of all x in i^{th} cluster, β_i = sum of all x^2 in i^{th} cluster.

Step 1: Read the first sample $\mathbf{x}_1$.
Create the first cluster around this sample and set the following.
$\mu_1 = \mathbf{x}_1$, $\sigma_1 = 0$, $c = 1$, $count_1 = 1$, $\alpha_i = \mathbf{x}$, $\beta_i = \mathbf{x}^2$
Step 2: Repeat the following steps until samples are available (or until not interrupted).
Step 3: Read the next sample $\mathbf{x}_i$.
Calculate the distance between $\mathbf{x}_i$ and all the existing cluster centres $\mathbf{cc}_j$
$dist_{ij} = \| \mathbf{x}_i - \mu_j \|$, for all $j = 1, ..., c$
Step 4: Select the cluster centres that satisfy the following:
$dist_{ij} > (max\ (\sigma_j\ , r) + r)$ for all $j = 1, ..., c$
Let $s1$ be the set of indices of all such cluster centres that satisfy the above condition.
Step 5: IF s1 is not empty THEN go to Step (6)
ELSE Create a new cluster around $\mathbf{x}_i$.
$c = c+1$, $\mu_c = \mathbf{x}_i$, $\sigma_c = 0$, $count_c = 1$, go to Step (2)
Step 6: Select the pth cluster centre that is closest to $\mathbf{x}_i$ and satisfies the condition given in Step (4).
$dist_{ip} = \| \mathbf{x}_i - \mu_p\| = \min(\|\mathbf{x}_i - \mu_1\|)$ for all $1 \in s1$
Considering that now $\mathbf{x}_i$ belongs to the pth cluster, update the cluster centre and average distance.
$\beta_p = \beta_p + \mathbf{x}_i^2$, $\alpha_p = \alpha_p + \mathbf{x}_i$

mean = $(count_p \times \mu_p + \mathbf{x}_i)$ / $(count_p + 1)$
variance = $(\beta_p + count_p \times \mathbf{mean}^2 - \times \mathbf{mean}^* \alpha_p)/(count_p + 1)$
μ_p = **mean**, σ_p = **variance**, $count_p = count_p + 1$
Step 7: Since now the position of the pth cluster centre has shifted.
Determine if it is required to be merged with any existing cluster centre that is close enough.
$dist_{pj} = \|\mu_p - \mu_j\|$ for all $j = 1, \ldots, c$ and $j \neq p$
Select the cluster centres that satisfy the following condition.
$dist_{pj} > max(\sigma_p, r) + max(\sigma_j, r)$ for all $j = 1, \ldots, c$ and $j \neq p$
Let $s2$ be the set of indices of all such cluster centres that satisfy the above condition.
Step 8: If $s2$ is not empty then select the closest cluster centre q.
$dist_{pq} = \| \mu_p - \mu_q \| = min\ (\|\mu_p - \mu_l\|)$ for all $l \in s2$
Merge cluster p and cluster q and update centre position, variance, and count.

B.4 *AutoCluster*

Algorithm 4 Pseudocode of the *AutoPredict* for prediction, estimation and filtration

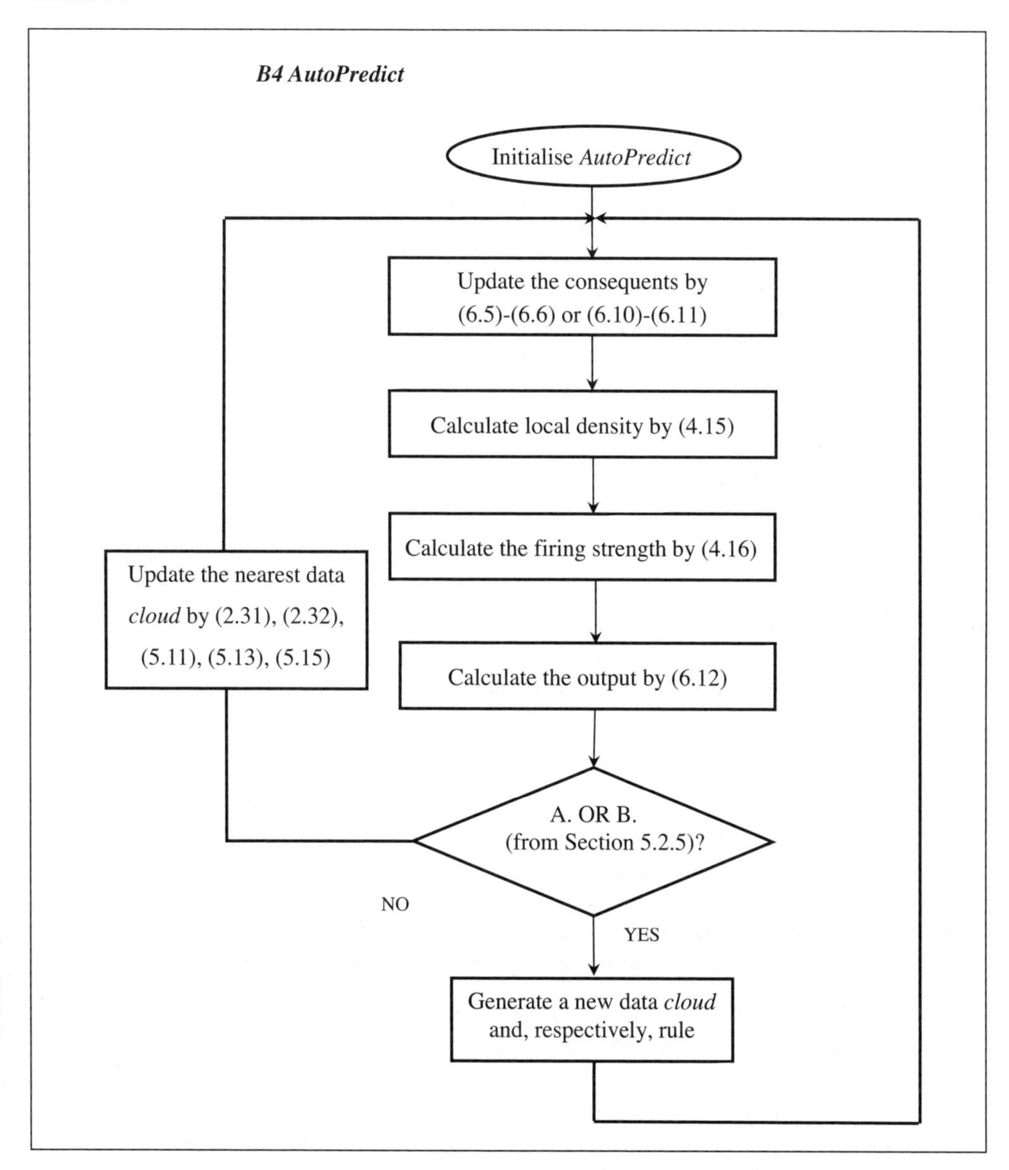

B.5 *AutoSense*

Algorithm 5 *AutoSensor*

```
Begin AutoSense
   Initialize AutoSense by the first data sample, z1 = [x1, y1];
   D1←1(or by iniSense if available); k←1.
    DO for each data sample WHILE data are acquired
     Read the measurable variables, xk;
     Calculate the density, D by RDE;
     Calculate the local submodel firing strength, λ;

    Estimate the outputs, ŷk;
    At the next time step (k ← k + 1)
    IF (mode='self-calibration')
         Get the real value of the estimated variables, yk;
         Apply the AutoPredict algorithm;
        END (self-calibration)
    END (DO...WHILE)
END (AutoSensor)
```

B.6 *AutoClassify0*

Algorithm 6 **Pseudocode of the** ***AutoClassify0***

```
Begin AutoClassify0
Initialize possibly 'from scratch'; IF 'from scratch' THEN
Form a focal point (and rule) based on the first data sample,
z1 = [x1, L1]; set its density to 1, D11←1; k←1.
Starting from the next data point
DO WHILE there is no more available data or
UNTIL a requirement to stop the process is received:

1) Read feature vector (inputs) of the data sample, xk;
2) Calculate the normalized firing strength, λk;
3) Determine the class label, L̂k;
4) At the next time step (k ← k + 1) get the true label, Lk;
5) Calculate the density, Dk(zk) in the input/output data space
   of the vector zk = [xk, Lk];
6) update the densities and quality parameters of the existing
   prototypes;

IF Principle A (see Section 5.2.5) holds
THEN
```

```
Add a new focal point based at the current data point;
Initiate its density to 1;
Update quality parameters for the focal points with labels
of the corresponding class.
IF Principle B (see Section 5.2.5) holds
THEN Remove the rules for which it holds;
END (IF)
ELSE
  Ignore (do not change the classifier structure);
  Update quality parameters for the focal points with labels of
  the corresponding class.
  Remove focal points with low support, utility and high Age.
 END (IF THEN ELSE)
END (DO...WHILE)
END (AutoClassify0)
```

B.7 *AutoClassify1*

Algorithm 7 Pseudocode of the *AutoClassify1* algorithm

```
Begin AutoClassify1
Initialize possibly 'from scratch'; IF 'from scratch' THEN
Form a focal point (and rule) based on the first data sample,
z1 = [x1, L1]; set its density to 1, D11 ← 1; k ← 1.
Starting from the next data point

DO WHILE there is no more available data or
UNTIL a requirement to stop the process is received:

1) Read feature vector (inputs) of the data sample, xk;
2) Calculate the normalized firing strength, λk;
3) Determine the class label, L̂k;
4) At the next time step (k ← k + 1) get the true label, Lk;
5) Calculate the density, Dk(zk) in the input/output data space
   of the vector zk = [xk, Lk];
6) update the densities and quality parameters of the exist-
   ing prototypes;

IF Principle A (see Section 5.2.5) holds
THEN
  Add a new focal point based on the current data point;
  Initiate its density to 1;
  Update quality parameters for the focal points with labels of
the corresponding class.
  Initialise the consequents parameters of local submodels.
```

```
   IF Principle B (see Section 5.2.5) holds
   THEN Remove the rules for which it holds;
   END (IF)

ELSE
    Ignore (do not change the classifier structure);
    Update quality parameters for the focal points with labels of
the corresponding class;
    Remove focal points with low support, utility and high Age;
    Update the consequent parameters of the local submodels.
    END (ELSE)
   END (DO..WHILE)
END (AutoClassify1)
```

B.8 *AutoControl*

Algorithm 8 Pseudocode of the *AutoControl* algorithm

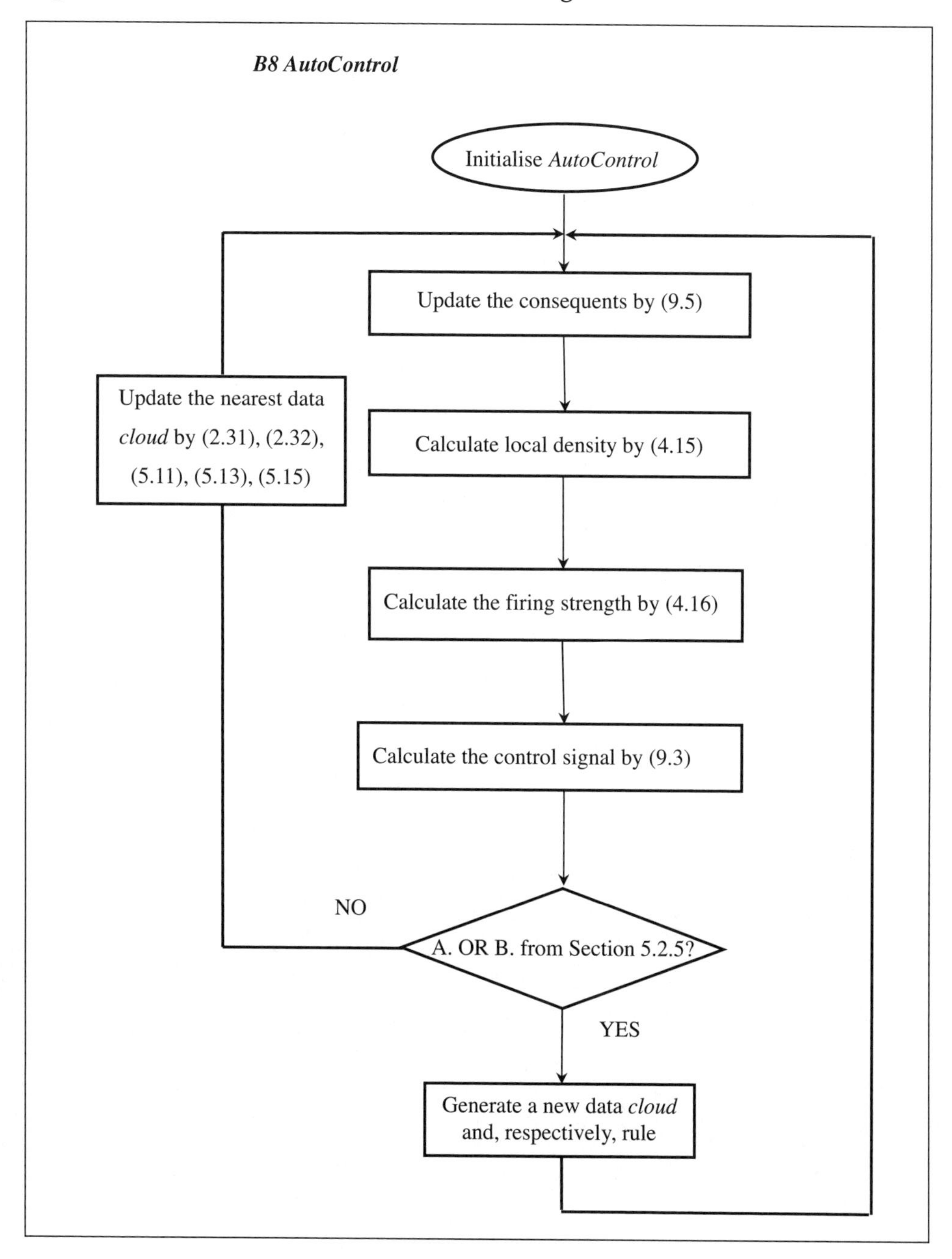

References

ActiveMedia (2004) Pioneer-3DX, User Guide, ActiveMedia Robotics, Amherst, NH, USA.

Akaike, H. (1970) Statistical predictor identification. *Annals of the Institute of Statistical Mathematics*, **22** (1), 203–217.

Angelov, P. and Tzonkov, S. (1993) Optimal Control of Biotechnological Processes Described by Fuzzy Sets. *Journal of Process Control*, **3** (3), 147–152.

Andersen, H.C., Teng, F.C. and Tsoi, A.C. (1994) Single net indirect learning architecture. *IEEE Transactions on Neural Networks*, **5** (6), 1003–1005.

Andreu, J., Angelov, P. and Dutta Baruah, R. (2011) *Real-time Recognition of Human Activities from Wearable Sensors by Evolving Classifiers*. IEEE International Conference on Fuzzy Systems (FUZZ-IEEE 2011), June 27–30, 2011, Taipei, Taiwan, pp. 2786–2793.

Angelov, P. (2002) *Evolving Rule-based Models: A Tool for Design of Flexible Adaptive Systems*, Springer Verlag, Heidelberg, Berlin, Germany.

Angelov, P. (2004a) An approach for fuzzy rule-base adaptation using on-line clustering. *International Journal of Approximate Reasoning*, **35** (3), 275–289.

Angelov, P. (2004b) A fuzzy controller with evolving structure. *Information Sciences*, **161** (1–2), 21–35.

Angelov, P. (2006) Machine Learning (Collaborative Systems) patent (WO2008053161, priority date: 1 Nov. 2006; international filing date 23 Oct. 2007); USA publication 11 Feb. 2010, # 2010–0036780, granted 21 August 2012.

Angelov, P. (2010) Evolving Takagi-Sugeno Fuzzy Systems from Data Streams (eTS+), In *Evolving Intelligent Systems: Methodology and Applications* (eds. P. Angelov, D. Filev and N. Kasabov), John Willey and Sons, IEEE Press Series on Computational Intelligence, pp. 21–50, ISBN: 978-0-470-28719-4, April 2010.

Angelov, P. (2011) *Autonomous Machine Learning (ALMA): Generating Rules from Data Streams*. Proc. Special International Conference on Complex Systems, COSY-2011, September 16–19, 2011, Ohrid, FYR of Macedonia, pp. 249–256.

Angelov, P. (2012) Anomalous System State Identification. Patent application GB1208542.9 priority date 15 May 2012.

Angelov, P., Baruah, R.D., Iglesias, J. and Andreu, J. (2011) STAKE: Real-time Spatio-Temporal Analysis and Knowledge Extraction through evolving clustering, Technical Report of the MoD project STAKE, DSTLX10000544446, June 30, 2011.

Autonomous Learning Systems: From Data Streams to Knowledge in Real-time, First Edition. Plamen Angelov.

Angelov, P. and Buswell, R. (2001) *Evolving Rule-based Models: A Tool for Intelligent Adaptation*. In Proc. of the Joint 9th IFSA World Congress and 20th NAFIPS Intern. Conf., July 25–28, 2001, Vancouver, BC, Canada, 2: pp. 1062–1066.

Angelov, P. and Buswell, R. (2002) Identification of evolving rule-based models. *IEEE Transactions on Fuzzy Systems*, **10** (5), 667–677.

Angelov, P. and Buswell, R. (2003) Automatic generation of fuzzy rule-based models from data by genetic algorithms. *Information Sciences*, **150** (1/2), 17–31.

Angelov, P. and Filev, D. (2004) An approach to on-line identification of evolving Takagi-Sugeno models. *IEEE Transactions on Systems, Man and Cybernetics, Part B Cybernetics*, **34** (1), 484–498.

Angelov, P. and Filev, D. (2005) *Simpl_eTS: A Simplified Method for Learning Evolving Takagi-Sugeno Fuzzy Models*, In Proc. Of The 2005 IEEE Intern. Conf. on Fuzzy Systems FUZZ-IEEE–2005, Reno, USA, pp. 1068–1073.

Angelov, P. and Filev, D. (2002) Flexible Models with Evolving Structure, IEEE Symposium on Intelligent Systems, Varna, Bulgaria, September 10–12, 2002, v.2, pp. 28–33.

Angelov, P. and Filev, D. (2003) *On-line Design of Takagi-Sugeno Models*, In: Lecture Notes in Computer Science 2715 Fuzzy Sets and Systems IFSA2003 (eds. T. Bilgiç, B. De Baets and O. Kaynak), pp. 576–584, ISBN 3-540-40383-3, 2003

Angelov, P., Kolev, D. and Markarian, G. (2012) Automatic System State Classification, patent application GB1218209.3, priority date 10 October 2012.

Angelov, P., Kordon, A. and Zhou, X. (2008) *Evolving Fuzzy Inferential Sensors for Process Industry, 3rd International Workshop on Genetic and Evolving Fuzzy Systems*, March 4–7, 2008, Witten-Bomerholz, Germany, pp. 41–46, ISBN 978-1-4244-1613-4.

Angelov, P. and Kordon, A. (2010) Adaptive inferential sensors based on evolving fuzzy models: an industrial case study. *IEEE Transactions on System, Man, and Cybernetics, Part B–Cybernetics, - Part B*, **40** (2), 529–539.

Angelov, P. and Yager, R. (2010) *A Simple Rule-based System through Vector Membership and Kernel-based Granulation*. Proc. 5th International Conference on Intelligent Systems, IS-2010, July 7–9, 2010, London, England, UK, IEEE Press, pp. 349–354.

Angelov, P. and Yager, R. (2011) A new type of simplified fuzzy rule-based systems. *International Journal of General Systems*, **41** (2), 163–185.

Angelov, P. and Yager, R. (2011) Simplified Fuzzy Rule-based Systems using Non-parametric Antecedents and relative Data Density, *IEEE Symposium Series on Computational Intelligence SSCI-2011*, April 11–15, 2011, Paris, France, pp. 62–69, ISBN 978-1-4244-9977-9.

Angelov, P. and Zhou, X. (2006) *Evolving Fuzzy Systems from Data Streams in Real-time. Proc. 2006 International Symposium on Evolving Fuzzy Systems*, Sept. 7–9, 2006, Ambleside, Lake District, UK, IEEE Press, pp. 29–35.

Angelov, P. and Zhou, X. (2007) Evolving fuzzy classifier for real-time novelty detection and landmark recognition by a mobile robot, in *Mobile Robots: The Evolutionary Approach* (eds N. Nedja, L. Coelho and L. Mourelle), *Studies in Computational Intelligence Series*, Springer, pp. 95–124.

Angelov, P. and Zhou, X. (2008) Evolving Fuzzy Rule-based Classifier from Data Streams. *IEEE Transactions On Fuzzy Systems*, **16** (6), 1462–1475.

Angelov, P., Andreu, J. and Vong, T. (2012) *Automatic Mobile Photographer and Assisted Picture Diary for Memory Aid*. IEEE Symposium on Evolving and Adaptive Intelligent Systems, EAIS-2012, May 17–18, 2012, Madrid, Spain.

Angelov, P., Buswell, R. and Wright, J.A. (2001) *Transparency and Simplification of Rule-Based Models for On-line Adaptation*. 2nd EUSFLAT Conference, Leicester, September 5–7, 2001, pp. 234–237.

Angelov, P., Ramezani, R. and Zhou, X. (2008) *Autonomous Novelty Detection and Object Tracking in Video Streams using Evolving Clustering and Takagi-Sugeno type Neuro-Fuzzy System*. 2008 IEEE International Joint Conference on Neural Networks, June 1–6, 2008, Hong Kong, pp. 1457–1464.

Angelov, I., Skrjanc, S. and Blazic, Self-learning Controllers, *In Springer Handbook on Computational Intelligence, (O. Castillo and P. Melin Ed.)*, 2012, to appear.

Angelov, P., Victor, J., Dourado A. and Filev, D. (2004a) *On-line evolution of Takagi-Sugeno Fuzzy Models*. Proc. 2nd IFAC Workshop on Advanced Fuzzy and Neural Control, Sept. 16–17, 2004, Oulu, Finland, pp. 67–72.

Angelov, P., Xydeas, C. and Filev, D. (2004b) *On-line Identification of MIMO Evolving Takagi-Sugeno Fuzzy Models*. Proc. of the Intern. Joint Conf. on Neural Networks and Intern. Conf. on Fuzzy Systems, IJCNN-FUZZ-IEEE, July 25–29, 2004, Budapest, Hungary, pp. 55–60.

Angelov, P., Zhou, X. and Klawonn, F. (2007a) *Evolving Fuzzy Rule-based Classifiers*. Proc. of the *First* 2007 *IEEE International Conference on Computational Intelligence Applications for Signal and Image Processing* – a part of the IEEE Symposium Series on Computational Intelligence, SSCI-2007, April 1–5, 2007, Honolulu, Hawaii, USA, pp. 220–225.

Angelov, P., Zhou, X., Lughofer, E. and Filev, D. (2007b) *Architectures of Evolving Fuzzy Rule-based Classifiers*. Proc. Of the 2007 IEEE International Conference on Systems, Man, and Cybernetics, Oct. 7–10, 2007, Montreal, Canada, pp. 2050–2055.

Angelov, P.P. and Kasabov, N. (2005) *Evolving Computational Intelligence Systems*. Proc. 1st Intern. Workshop on Genetic Fuzzy Systems, Granada, Spain, pp. 76–82.

Ankenbrandt, C.A., Buckles, B.P. and Petrya, F.E. (1990) Scene recognition using genetic algorithms with semantic nets. *Pattern Recognition Letters*, **11** (4), 285–293.

ARIA (2011) http://robots.mobilerobots.com/ARIA/README.txt (last accessed November 2011).

Arulampalam, M., Maskell, S. and Gordon, N. (2002) A tutorial on particle filters for on-line nonlinear non-Gaussian Bayesian tracking. *IEEE Transactions on Signal Processing*, **50** (2), 174–188.

ASTM (2011) American Standard Test Method for Distillation of Petroleum Products at Atmospheric Pressure, http://myastm.astm.org/filtrexx40.cgi?-P+MEM_NUM++ P+cart++/usr6/htdocs/newpilot.com/subscription/historical/D86-04B.htm (last accessed November 2011).

Astroem, K. and Wittenmark, B. (1989) *Adaptive Control*, Addison Wesley, Massachusetts, USA.

Azimi-Sadjadi, M., Yao, D., Jamshidi, A. and Dobeck, G. (2002) Underwater target classification in changing environments using an adaptive feature mapping. *IEEE Transactions on Neural Networks*, **13** (5), 1099–1111.

Azimov, I. (1950) *I, Robot*, Doubleday & Company, New York.

Babuska, R. (1998) *Fuzzy Modelling*, Kluwer Academic Publishers, Norwell, MA, USA.

Badami, V. and Chbat, N. (1998) Home appliances get smart. *IEEE Spectrum*, **35** (8), 36–43.

Baldwin, J.F., Martin, T.P. and Pilsworth, B.W. (1995) *Fril-fuzzy Evidential Reasoning in Artificial Intelligence*, Published by Research Studies Press (RSP), Taunton, UK. Marketed by John Wiley & Sons Ltd, Baffins Lane, Chichester, West Sussex, PO19 1UD, UK. Tel. +44 243 779777, Fax. +44 243 775878. Registered No: 641132, England. ISBN 0 86380 159 5, 404 pp, published Jan. 1995.

Bar-Shalom, Y., Rong, L.X. and Kirubarajan, T. (2001) *Estimation with Applications to Tracking and Navigation: Theory, Algorithms and Software*, John Wiley & Sons, Inc. Wiley Interscience Publication, New York, 2001.

Baruah, R. and Angelov, P. (2010) *Clustering as a Tool for Self-generation of Intelligent Systems: A Survey*. Proc. International Conference on Evolving Intelligent Systems, EIS'10, Leicester, UK, pp. 34–41.

Baruah, R. and Angelov, P. (2012) *A New On-line Clustering Algorithm: Evolving Local Means*. Proc. 2012 IEEE World Congress on Computational Intelligence, June 10–15, 2012, Brisbane, Australia, pp. 2161–2168 (IEEE Press ISBN 978-1-4673-1489-3).

Berenji, H.R. and Khedkar, P.S. (1993) *Adaptive Fuzzy Control with Reinforcement Learning*. American Control Conference, June 1993, San Francisco, CA.

Berry, E. *et al.* (2007) The use of a wearable camera, SenseCam, as a pictorial diary to improve autobiographical memory in a patient with limbic encephalitis: a preliminary report. *Neuropsychological Rehabilitation*, **17**, 582–601.

Bezdek, J. (1974) Cluster validity with fuzzy sets. *Journal of Cybernetics*, **3** (3), 58–71.

Bharitkar, S. and Filev, D. (2001) An online learning vector quantization algorithm. In *Proc. of Sixth International Symposium on Signal Processing and its Applications*, 2: pp. 394–397.

Bishop, C. (2009) *Pattern Recognition and Machine Learning*, 2nd edn., Springer, NY, USA.

Boukhris, A., Mourot, G. and Ragot, J. (1999) Non-linear dynamic system identification: a multi-model approach. *International Journal of Control*, **72** (7–8), 591–604.

Buehler, M., Iagnemma, K. and Singh, S. (eds) (2010) Autonomous vehicles in city traffic, *The DARPA Urban Challenge, Series: Springer Tracts in Advanced Robotics*, Springer-Verlag, Berlin, Heidelberg, 2009, 56, pp. 626.

Cara, A., Lendek, Z., Babuska, R. *et al.* (2010) *Online Self-Organizing Adaptive Fuzzy Controller: Application to a Nonlinear Servo System*. International Conference on Information Processing and Management of Uncertainty in Knowledge-Based Systems.

Carline, D., Angelov, P. and Clifford, R. (2005) *Agile Collaborative Autonomous Agents for Robust Underwater Classification Scenarios*. In the Proceedings of the Underwater Defense Technology Conference, *June 2005, Amsterdam*.

Carpenter, G. and Grossberg, S. (2003) Adaptive resonance theory, in *The Handbook of Brain Theory and Neural Networks*, 2nd edn. (ed. M.A. Arbib), MIT Press, Cambridge, MA, USA, pp. 87–90.

Carse, B., Fogarty, T. and Munro, A. (1996) Evolving fuzzy rule-based controllers using GA. *Fuzzy Sets and Systems*, **80**, 273–294.

CASAS (2010) *Smart home project, School of Electrical Engineering and Computer Science – Washington State University*, available on-line at http://ailab.wsu.edu/casas/ (last accessed 12 February 2012).

Chen, L., Nguang, S., Li, X.M. and Chen, X. (2004) Soft sensors for on-line biomass measurements. *Bioprocess Biosystem Engineering*, **26** (3), 191–195.

Cheung, S.-C. and Kamath, C. (2004) *Robust Techniques for Background Subtraction in Urban Traffic Video*. In Proc. SPIE, Electronic Imaging Video Comm. and Image Proc., San Jose, pp. 881–892.

Chiu, S. (1994) Fuzzy model identification based on cluster estimation. *Journal of Intelligent and Fuzzy Systems*, **2**, 267–278.

Choset, H. and Nagatani, K. (2001) Topological simultaneous localization and mapping (SLAM): toward exact localization without explicit localization. *IEEE Transactions on Robotics and Automation*, **17** (2), 125–137.

Christiani, M., Farenzena, M., D. Bloisi and V. Murino (2010) Background subtraction for automated multisensory surveillance: A comprehensive review. *Journal of Advances in Signal Processing*.

Comaniciu, D. and Meer, P. (2002) Mean shift: a robust approach toward feature space analysis. *IEEE Transactions on Pattern Analysis and Machine Intelligence*, **24** (5), 603–619.

Cordon, O., Gomide, F., Herrera, F. *et al.* (2004) Ten years of genetic fuzzy systems: current framework and new trends. *Fuzzy Sets and Systems*, **141** (1), 5–31.

Dagher, I. (2010) *Incremental PCA-LDA Algorithm*. International Conference on Computational Intelligence for Measurement Systems and Applications, pp. 97–101.

DailyMail (2009) http://www.dailymail.co.uk/news/article-1205607/Shock-figures-reveal-Britain-CCTV-camera-14-people–China.html (last accessed December 2011).

Demspter, A.P. (1968) A generalization of Bayesian inference. *Journal of Royal Statistics Society*, **30**, 205–247.

Desai, J., Ostrowski, J. and Kumar, V. (2001) Modeling and control of formation of nonholonomic mobile robots. *IEEE Transactions Robot Automation*, **17**, 905–908.

Detyenecki, M. and Tateson, R. (2005) *Nature-Inspired Networks: The Telecommunications Industry Point of View*, Activity Report, BT, NiSiS.

Domingos, P. and Hulten, G. (2001) *Catching up with the data: Research issues in mining data streams*. Workshop on Research Issues in Data Mining and Knowledge Discovery, Santa Barbara, CA.

Doucet, A., Godsill, S. and Andrieu, C. (2000) On sequential Monte Carlo sampling methods for Bayesian filtering. *Statistics and Computing*, **10** (3), 197–208.

Driankov, D., Hellendoorn, H. and Reinfrank, M. (1993) *An Introduction to Fuzzy Control*, Springer, Verlag, Berlin, Heidelberg, New York.

Dubois, D., Prade, H. and Lang, J. (1990) *Fuzzy Sets and Approximate Reasoning*, IRIT.

Duda, R.O., Hart, P.E. and Stork, D.G. (2000) *Pattern Classification*, John Wiley & Sons, NY, USA.

Elgammal, A., Duraiswami, R., Harwood, D. and Davis, L. (2002) Background and Foreground Modelling Using Non- Parametric Kernel Density Estimation for Visual Surveillance KDE. *Proceedings of the IEEE* In Proceedings of the IEEE, Vol. 90, No. 7. (July 2002), pp. 1151–1163.

Endsley, M. (1996) Automation and situational awareness, in *Automation and Human Performance: Theory and Applications* (eds R. Parasuraman and M. Mouloua), Earlbaum, Mahwah, New Jersey, pp. 163–181.

Everett, M. and Angelov, P. (2005) *EvoMap: On-Chip Implementation of Intelligent Information Modelling using Evolving Mapping*, Lancaster University, pp. 15.

Fayyad, U., Piatetsky-Shapiro, G. and Smyth, P. (1996) *From Data Mining to Knowledge Discovery: An Overview, Advances in Knowledge Discovery and Data Mining*, MIT Press, Massachusetts, USA.

Ferreyra, A. and Rubio, J. (2006) *A New On-line Self-constructing Neural Fuzzy Network*. Proc. 45th IEEE Conf. Decision Control, Dec. 13–15, pp. 3003–3009.

Filev, D. and Kolmanovsky, I. (2012) Markov chain modeling and on-board identification for automotive vehicles identification for automotive systems, in *Lecture Notes in Control and Information Sciences*, Editors: Daniel Alberer, Håkan Hjalmarsson, Luigi de lRe; Identification for Automotive Systems, Lecture Notes in Control and Information Sciences, ISBN: 978-1-4471-2220-3 (Print) 978-1-4471-2221-0 (Online), vol. **418**, Springer, Berlin/Heidelberg, pp. 111–128.

Filev, D. and Tseng, F. (2006) *Novelty Detection-based Machine Health Prognostics*, In the Proc. of the 2006 Intern. Symposium on Evolving Fuzzy Systems, IEEE Press, pp. 193–199.

Filev, D., Larson, T. and Ma, L. (2000) *Intelligent Control for Automotive Manufacturing – Rule-based Guided Adaptation*. Proc. of the IEEE Conference on Industrial Electronics, IECON-2000, Oct. 2000, Nagoya, Japan, pp. 283–288.

Fisher, R. (1936) The use of multiple measurements in taxonomic problems. *Annals of Eugenics*, **7** (2), 179–188.

Fortuna, L., Graziani, S., Rizzo, A. and Xibilia, M. (2007) *Soft sensors for Monitoring and Control of In Industrial Processes*, Springer-Verlag, London, UK.

Fredkin, E. (1960) Trie memory. *Communications of the ACM*, **3** (9), 490–499.

Fritzke, B. (1994) Growing cell structures–a self-organizing network for unsupervised and supervised learning. *Neural Networks*, **7** (9), 1441–1460.

Fukunaga, K. and Hostetler, L.D. (1975) The estimation of the gradient of a density function, with applications in pattern recognition. *IEEE Transactions on Information Theory*, **21**, 32–40.

Fung, G. and Mangasariany, O. (2002) *Incremental Support Vector Machine Classifier*. Proceedings of the Second SIAM International.

Gabor, D. (1946) Theory of communications. *Proceedings Institute of Electrical Engineering*, **93**, 429–459.

Gao, Y. and Er, M.J. (2003) Online adaptive fuzzy neural identification and *control* of a class of MIMO nonlinear systems. *IEEE Transactions on Fuzzy Systems*, **11** (4), 12–32.

Georgieva, O. and Filev, D. (2010) An extended version of the Gustafson-Kessel algorithm for evolving data stream clustering, in *Evolving Intelligent Systems: Methodology and Applications*, eds. P. Angelov, D. Filev and N. Kasabov John Wiley and Sons, Hoboken, NJ, USA, pp. 273–300.

Giarratano, J. and Riley, G. (1998) *Expert Systems*, PWS Publishing Co., Boston, MA, USA.

Godoy, D. and Amandi, A. (2005) User profiling for web page filtering. *IEEE Internet Computing*, **9** (4), 56–64.

Goldberg, D. (1989) *Genetic Algorithms in Search, Optimization and Machine Learning*, Addison-Wesley, Reading, MA, USA.

Greenberg, S. (1988) Using UNIX: Collected traces of 168 users. Master's thesis, Department of Computer Science, University of Calgary, Alberta, Canada.

Gustafson, D.E. and Kessel, W.C. (1978) *Fuzzy Clustering with a Fuzzy Covariance Matrix*, Proc. IEEE Conference on Decision and Control, including the 17th Symposium on Adaptive Processes, Jan. 1978, pp. 761–766.

Hampapur, A. (2005) Smart video surveillance, exploring the concept of multi-scale spatiotemporal tracking. *IEEE Signal Processing Magazine*, **22** (2), 38–51.

Han, K. and Veloso, M. (1999) *Automated Robot Behavior Recognition Applied to Robotic Soccer*. In Proc. IJCAI-99 Workshop on Team Behaviors and Plan Recognition.

Harris, C.J. (ed.) (1994) *Advances in Intelligent Control*. Taylor and Francis, London, UK.

Hastie, T., Tibshirani, R. and Friedman, J. (2001) *The Elements of Statistical Learning: Data Mining, Inference and Prediction*, Springer Verlag: Heidelberg, Germany.

Haykin, S. (2002) *Adaptive Filter Theory*, 4th edn., Prentice Hall, Upper Saddle River NJ, USA.

Healy, M., Newe, T. and Lewis, E. (2008) *Wireless Sensor Node Hardware: A Review*. In 7th IEEE Conference on Sensors, Lecce, Italy.

Hernandez, J.M. and Angelov, P. (2010) Applications of Evolving Intelligent Systems to Oil and Gas Industry, In *Evolving Intelligent Systems: Methodology and Applications* (eds. P. Angelov, D. Filev and N. Kasabov), John Willey and Sons, IEEE Press Series on Computational Intelligence, pp. 399–420, ISBN: 978-0-470-28719-4, April 2010.

Hill, A., Crazer, F. and Wilkinson, P. (2007) *Effective Operator Engagement with Variable Autonomy*. Proc. 2nd SEAS-DTC Technical Conference, B5, June 2007, Edinburgh, Scotland, UK.

Hodges, S., Williams, L., Berry, E. *et al.* (2006) SenseCam: a retrospective memory aid. in *UbiComp 2006: Ubiquitous Computing*, vol. **4206** (eds P. Dourish and A. Friday), Springer, Berlin/Heidelberg, pp. 177–193.

Holland, J.H. (1975) *Adaptation in Natural and Artificial Systems: An Introductory Analysis with Applications to Biology, Control and Artificial Intelligence (Complex Adaptive Systems)*, MIT Press: Boston, MA, USA.

Hopner, F. and Klawonn, F. (2000) *Obtaining Interpretable Fuzzy Models from Fuzzy Clustering and Fuzzy Regression*. In Proc. of the 4th Intern. Conf. on Knowledge-based Intelligent Engineering Systems (KES), Brighton, UK, pp. 162–165.

Horak, O. (1993) Standardized methods of testing flammability of plastics. *Makromolekulare Chemie, Macromolecular Symposia*, **74**, 339–342.

Hornby, A. (1974) *Oxford Advance Learner's Dictionary*, Oxford University Press.

Hotteling, H. (1931) The generalisation of Student's ratio. *Annals of Mathematical Statistics*, **2** (3), 360–378.

IBM strategy on Autonomic Computing (2009) http://www.research.ibm.com/autonomic/overview/solution.html, visited April 2009.

Iglesias, J.A., Angelov, P., Ledezema, A. and Sanchis, A. (2009) *Modelling Evolving User Behaviours*. Proc. 2009 IEEE Symposium Series on Computational Intelligence, 2 April, 2009, Nashville, TN, USA, pp. 16–23.

Iglesias, J.A., Angelov, P., Ledezma, A. and Sanchis, A. (2010) Human activity recognition based on evolving fuzzy systems. *International Journal of Neural Systems*, **20** (5), 355–364.

Iglesias, J.A., Angelov, P., Ledezma, A. and Sanchis, A. (2012) Creating user behaviour profiles automatically. *IEEE Transactions on Knowledge and Data Engineering*, **24** (5), 854–867.

Ishibuchi, H., Nakashima, T. and Nii, M. (2004) *Classification and Modeling with Linguistic Granules: Advanced Information Processing*, Springer Verlag, Berlin.

Ishida, H. and Iwama, A. (1984) Ignition characteristics of gelled (0/W Emulsified) hydrocarbon fuel pool. *Combustion Science and Technology*, **36** (1–2), 51–64.

Jang, J.S.R. (1993) ANFIS: adaptive network-based fuzzy inference systems. *IEEE Transactions on Systems, Man & Cybernetics, Part B–Cybernetics*, **23** (3), 665–685.

Juang, C.-F. and Lin, C.-T. (1999) A recurrent self-organizing neural fuzzy inference network. *IEEE Transactions on Neural Networks*, **10** (4), 828–845.

Kacprzyk, J. and Zadeh, L.A. (1999) *Computing with Words in Information/intelligent Systems: Foundations*, Physica-Verlag, Heidelberg, New York.

Kailath, T., Sayed, A.H. and Hassibi, B. (2000) *Linear Estimation*, Prentice Hall, Upper Saddle Rive, NJ, USA.

Kalman, R.E. (1960) A new approach to linear filtering and prediction problem. *Transactions of the American Society of Mechanical Engineering, ASME, Ser. D, Journal of Basic Engineering*, **82** (Serise D), 34–45.

Kanakakis, V., Valavanis, K.P. and Tsourveloudis, N.C. (2004) Fuzzy-logic based navigation of underwater vehicles. *Journal of Intelligent and Robotic Systems*, **40** (1), 45–88.

Karnik, N.N., Mendel, J.M. and Liang, Q. (1999) Type-2 fuzzy logic system. *IEEE Transactions on Fuzzy Systems*, **7** (6), 643–658.

Kasabov, N. (2001) Evolving fuzzy neural networks for on-line supervised/unsupervised, knowledge-based learning. *IEEE Transactions on Systems, Man and Cybernetics – Part B, Cybernetics*, **31** (6), 902–918.

Kasabov, N. (2002) Evolving connectionist systems for adaptive learning and knowledge discovery: method, tools, applications. *IEEE International Conference on Intelligent Systems*, **1**, 24–28.

Kasabov, N. (2006a) Adaptation and interaction in dynamical systems: modelling and rule discovery through evolving connectionist systems. *Applied Soft Computing*, **6** (3), 307–322.

Kasabov, N. (2006b) *Evolving Connectionist Systems: Brain-, Gene-, and, Quantum Inspired Computational Intelligence*, Springer Verlag, London, Heidelberg, NY.

Kasabov, N. and Filev, D. (2006) *Evolving Intelligence Systems: Methods, Learning, & Applications*. International Symposium on Evolving Fuzzy Systems, Sept. 7–9, 2006, Ambleside, Lake District, plenary talk.

Kasabov, N. and Song, Q. (2002) DENFIS: dynamic evolving neural-fuzzy inference system and its application for time-series prediction. *IEEE Transactions on Fuzzy Systems*, **10** (2), 144–154.

Kasabov, N.K. (1998) Evolving fuzzy neural networks: theory and applications for on-line adaptive prediction, decision making and control. *Australian Journal on Intelligent Information Processing Systems*, **5** (3), 154–160.

Kelly, J. *et al.* (2008) A self-learning fuzzy classifier with feature selection for intelligent interrogation of mid-IR spectroscopy data derived from different categories of exfoliative cervical cytology. *International Journal on Computational Intelligence Research*, **4** (4), 392–401.

Kelly, J.G. *et al.* (2010) Robust classification of low-grade cervical cytology following analysis with ATR-FTIR spectroscopy and subsequent application of self-learning classifier eClass. *Journal of Analytical and Bio-analytical Chemistry*, **398** (5), 2191–2201.

Kephart, J.O. (1994) *A Biologically Inspired Immune System for Computers*. Proc. Artificial Life IV: The Fourth Intern. Workshop on the Synthesis and Simulation of Living Systems, MIT Press, pp. 130–139.

Kleeman, L. (1992) Optimal estimation of position and heading for mobile robots using ultrasonic beacons and dead-reckoning. *IEEE Transactions on Robotics and Automation*, **3**, 2582–2587.

Klinkenberg, R. and Joachims, T. (2000) *Detection Concept Drift with Support Vector Machines*. Proc. of the 7th International Conference on Machine Learning (ICML), Morgan Kaufman, pp. 487–494.

Kohonen, T. (1982) Self-organizing formation of topologically correct feature maps. *Biological Cybernetics*, **43** (1), 59–69.

Kohonen, T. (1984) *Self-organisation and Association Memory*, Springer-Verlag, New York.

Kohonen, T. (1995) *Self-Organizing Maps*, Series in Information Sciences, Vol. 30. Springer Verlag, Heidelberg, Germany.

Kolev, D. *et al.* (2012a) Autonomous flight data analysis by recursive density estimation and statistical analysis methods. *IEEE Transactions on Intelligent Transport*, submitted.

Kolev, D. *et al.* (2012b) A non-symmetrical single-class SVM classifier for flight data analysis. *IEEE Transactions on Neural Networks*, submitted

Kordon, A. (2006) *Inferential Sensors as Potential Application Area of Intelligent Evolving Systems*. International Symposium on Evolving Fuzzy Systems, Ambleside, Lake District, UK, key note presentation.

Kordon, A. and Smits, G. (2001) *Soft Sensor Development using Genetic Pro Gramming*. Proc. GECCO, San Francisco, CA, pp. 1346–1351.

Kordon, A. *et al.* (2003) Robust soft sensor development using GP, in *Nature-Inspired Methods in Chemometrics* (ed. R. Leardi), Elsevier, Amsterdam, The Netherlands, pp. 69–108.

Kovacs, T. and Bull, L. (2005) *Foundation of Learning Classifier Systems (Studies in Fuzziness and Soft Computing)*, Springer-Verlag.

Koza, J. (1992) *Genetic Programming: On the Programming of Computers by Means of Natural Selection*, MIT Press, USA.

Krishnakumar, K. (2003) Artificial Immune System Approaches for Aerospace Applications. American Institute of Aeronautics and Astronautics, 41st Aerospace Science Meeting and Exhibit, 6–9 Jan. 2003, Reno, Nevada, USA.

Kuncheva, L. (2000) *Fuzzy Classifiers*, Physica-Verlag, Heidelberg, Germany.

Leng, G., McGuinty, T.M. and Prasad, G. (2005) An approach for on-line extraction of fuzzy rules using a self-organizing fuzzy neural network. *Fuzzy Sets and Systems*, **150** (2), 211–243.

Li, W., Yue, H.H. and Valle-Cervanteset, S. (2000) Recursive PCA for adaptive process monitoring. *Journal of Process Control*, **10** (5), 471–486.

Li, X., Wu, C., Zach, C. *et al.* (2008) *Modeling and Recognition of Landmark Image Collections Using Iconic Scene Graphs*. Proc. European Conference on Computer Vision, ECCV08, 12–18 October 2008, Marseille, France; LNCS, Part I, pp. 427–440, Springer Verlag, Berlin-Heidelberg, Germany.

Lin, F.J., Lin, C.H. and Shen, P.H. (2001) Self-constructing fuzzy neural network speed controller for permanent-magnet synchronous motor drives. *IEEE Transactions on Fuzzy Systems*, **9** (5), 751–759.

Liu, J. (2005) *On-line Soft Sensor for Polyethylene Process with Multiple Production Grades*. Proc. 16th IFAC World Congr., Prague, Czech Republic.

Liu, L. and Yager, R.R. (2008) *Classic Works of the Dempster-Shafer Theory of Belief Functions: An Introduction, Studies in Fuzziness and Soft Computing*, Springer, Vol. 219/2008, 1–34.

Liu, P.X. and Meng, M.Q-X. (2004) On-line data-driven fuzzy clustering with applications to real-time robotic tracking. *IEEE Transactions on Fuzzy Systems*, **12** (4), 516–523.

Ljung, L. (1987) *System Identification: Theory for the User*, Prentice-Hall, New Jersey, USA.

Lughofer, E. (2011) *Evolving Fuzzy Systems – Methodologies, Advanced Concepts and Applications*, Physica-Verlag, Heidelberg, Berlin.

Lughofer, E. and Angelov, P. (2011) Handling drifts and shifts in on-line data streams with evolving fuzzy systems. *Applied Soft Computing*, Elsevier, **11** (2), 2057–2068.

Lughofer, E., Angelov, P. and Zhou, X. (2007) *Evolving Single- and Multi-Model Fuzzy Classifiers with FLEXFIS-Class*. Proc. of the 2007 IEEE International Conference on Fuzzy Systems, 23–26 July, 2007, London, UK, pp. 363–368.

Macias, J. and Feliu, J.A. (2001) *Dynamic Study of Inferential Sensors (NN) in Quality Prediction of Crude Oil Distillation Tower Side Streams*. 11th European Symposium on Computer Aided Process Engineering, ESCAPE 11, May 2001, Kolding, Denmark.

Macias, J., Angelov, P. and Zhou, X.-W. (2006) *Predicting Quality of the Crude Oil Distillation using Evolving Takagi-Sugeno Fuzzy Models*, In Proc. 2006 International Symposium on Evolving Fuzzy Systems, September 7–9, 2006, Ambelside, Lake District, UK, IEEE Press, pp. 201–207, ISBN 0-7803-9719-3.

Macias-Hernandez, J.J., Angelov, P. and Zhou X. (2007) *Soft Sensor for Predicting Crude Oil Distillation Side Streams using Takagi–Sugeno Evolving Fuzzy Models*. Proc. 2007 IEEE Intern. Conference on Systems, Man, and Cybernetics, 7–10 Oct., 2007, Montreal, Canada, pp. 3305–3310.

Mamdani, E.H. and Assilian, S. (1975) An experiment in linguistic synthesis with a fuzzy logic controller. *International Journal of Man-Machine Studies*, **7**, 1–13.

Manevitz, L.M. and Yousef, M. (2001) One-class SVMs for document classification. *Journal of Machine Learning Research*, **2**, 139–154.

Marin-Blazquez, J.G. and Shen, Q. (2002) From approximative to descriptive fuzzy classifiers. *IEEE Transaction on Fuzzy Systems*, **10** (4), 484–497.

Martin, T. (2005) Fuzzy Sets in the fight against digital obesity. *Fuzzy Sets and Systems*, **156** (3), 411–417.

McDonald, S., Xydeas, C. and Angelov, P. (2008) *Decision Support Systems – Improving levels of Care and Lowering the Costs in Anticoagulation Therapy*. First International Conference on Electronic Healthcare for the 21st Century, eHelath, London, UK, pp. 175–178.

Michalewicz, Z. (1996) *Genetic Algorithms + Data Structures = Evolution Programs*, 3rd edn., Springer-Verlag, Belin, Heidelberg, New York.

Murillo, A.C. and Kosecka, J. (2009) *Experiments in Place Recognition using Gist Panoramas*. 9th IEEE Workshop on Omnidirectional Vision, Camera Networks and Non-classical Cameras (OMNIVIS), with Int. Conf. on Computer Vision, pp. 2196–2203.

Murphy, K., Torralba, A. and Freeman, W. (2004) Using the forest to see the trees: a graphical model relating features, objects, and scenes. *Advances in Neural Information Processing Systems*, **16**, 1499–1506.

Naisbitt, J. (1988) *Megatrends: Ten New Directions Transforming Our Lives*, Grand Central Publ.

Narendra, K.S. and Parthasarathy, K. (1990) Identification and control of dynamical systems using neural networks, *IEEE Transactions on Neural Networks*, **1** (1), 4–27.

Nehmzow, U., Smithers, T. and Hallam, J. (1991) Location recognition in a mobile robot using self-organizing feature maps, in *Information Processing in Autonomous Mobile Robots* (ed. G. Schmidt), Springer, Berlin.

Netto, H.V. (2006) *Visual Novelty Detection for Autonomous Inspection Robots*, PhD Thesis, University of Essex, Colchester, UK.

Newman, J.R. (1956) *The World of Mathematics*, Simon and Schuster, New York.

Oliva, A. and Torralba, A. (2002) *Scene-Centered Description from Spatial Envelope Properties*. Proc. International Workshop on Biologically Motivate Computer Vision, BMCV 2002, 22–24 November 2002, Tubingen, Germany (eds H.H. Bulthoff, *et al.*), LNCS 2525, pp. 263–272, Springer-Verlag, Berlin-Heidelberg, Germany.

Oliva, A. and Torralba, A. (2001) Modeling the shape of the scene: a holistic representation of the Spatial Envelope. *International Journal of Computer Vision*, **42**, 145–175.

Oliva, A. and Torralba, A. (2006) Building the gist of a scene: the role of global image features in recognition, *Visual Perception, Progress in Brain Research*, **155**, 23–36.

de Oliveira, J.V. (1999) Semantic constraints for membership function optimization. *IEEE Transactions on Systems, Man and Cybernetics, Part A: Systems and Humans*, **29** (1), 128–138.

Ozawa, S., Pang, S. and Kasabov, N. (2005) A Modified Incremental Principal Component Analysis for On-Line Learning of Feature Space and Classifier, *Lecture Notes in Artificial Intelligence* LNAI, Vol. 3157, Springer-Verlag: Berlin, Heidelberg, Germany, pp. 231–240.

Pang, S., Ozawa, S. and Kasabov, N. (2005) Incremental linear discriminant analysis for classification of data streams. *IEEE Transactions on Systems, Man and Cybernetics, Part B – Cybernetics*, **35** (5), 905–914.

Papoulis, A. (1991) *Probability, Random Variables, and Stochastic Processes*, 3rd edn., McGraw-Hill, pp. 113–114.

Patton, R.J. *et al.* (2000) *Issues of Fault Diagnosis for Dynamic Systems*, Springer-Verlag, London, pp. 87–114.

Pedrycz, W. (1993) *Fuzzy Control and Fuzzy Systems*, 2nd edn., Research Studies Press Ltd., Taunton, UK.

Pedrycz, W. (1994) Why triangular membership functions?. *Fuzzy Sets and Systems*, **64** (1), 21–30.

Pepyne, D.L., Hu, J. and Gong, W. (2004) User profiling for computer security, In *Proc. American Control Conference*, pp. 982–987.

Platt, J. (1991) A resource allocation network for function interpolation. *Neural Computation*, **3** (2), 213–225.

Poirier, F. and Ferrieux, A. (1991) DVQ: Dynamic vector quantization – An incremental LVQ. *Proceeding International Conference Artificial Neural Networks*, **2**, 1333–1336.

Porikli, F. and Tuzel, O. (2003) Human body tracking by adaptive background and mean-shift analysis, *IEEE International Workshop on Performance Evaluation of Tracking and Surveillance*.

Procyk, T.J. and Mamdani, E.H. (1975) A linguistic self-organizing process controller, *Automatica*, **15**, 15–30.

Psaltis, D., Sideris, A. and Yamamura, A.A. (1988) A multilayered neural network controller. *IEEE Control Systems Magazine*, **8**, 17–21.

Qin, S.J., Yue, H. and Dunia, R. (1997) Self-validating inferential sensors with application to air emission monitoring. *Industrial Engineering Chemistry Research*, **36**, 1675–1685.

Quinlan, J. (2003) Data Mining tools see5 and C5.0., available on-line at http://www.rulequest.com/see5-info.html (last accessed 19 January 2012).

Rabiner, L.R. (1989) A Tutorial on hidden Markov models and selected applications in speech recognition. *In Proceeding of the IEEE*. **77** (2), 257–286.

Ramezani, R., Angelov, P. and Zhou, X. (2008) A Fast Approach to Novelty Detection in Video Streams using Recursive Density Estimation, *4th International IEEE Symposium on Intelligent Systems*, September 6–8, 2008, Varna, Bulgaria, v. II, pp. 14-2–14-7, ISBN 978-1-4244-1739-1/08.

Ren, W. and Beard, R.W. (2003) *A Decentralized Scheme for Spacecraft Formation Flying via the Virtual Structure Approach*. Proceeding of the American Control Conference, pp. 1746–1751.

Rittscher, J. *et al.* (2000) *A Probabilistic Background Subtraction Model for Tracking*. Proc. IEEE Conference on Computer Vision.

Rumelhart, D.E., McClelland, J.L., the PDP Research Group (1986) *Parallel Distributed Processing: Explorations in the Microstructure of Cognition. Volume 1: Foundations*, MIT Press, Cambridge, MA.

Sadeghi-Tehran, P. *et al.* (2012) *Self-evolving Parameter-free Rule-based Controller*. World Congress on Computational Intelligence, 10–15 June 2012, Brisbane, Australia, pp. 754–761 (IEEE Press, ISBN 978-1-4673-1489-3).

Sadeghi-Tehran, P. and Angelov, P. (2012) A Real-time Approach for Novelty Detection and Trajectories Analysis for Anomaly Recognition in Video Surveillance Systems, *Proc. 2012 IEEE Conference on Evolving and Adaptive Intelligent Systems*, Madrid, May 16–17, 2012, pp. 108–113, ISBN 978-1-4673-1727-6/12.

Sadeghi-Tehran, P., Angelov, P. and Behera, S. (2011) Autonomous Visual Self-localization in Completely Unknown Environment. *Proc. 2012 IEEE Conference on Evolving and Adaptive Intelligent Systems*, Madrid, 16–17 May 2012, pp. 90–95, ISBN 978-1-4673-1727-6/12.

Sadeghi-Tehran, P., Angelov, P. and Ramezani, R. (2010) A Fast Approach to Autonomous Detection, Identification, and Tracking of Multiple Objects in Video Streams under Uncertainties, In: (eds. E. Hüllermeier, R. Kruse and F. Hoffmann): IPMU 2010, Part II, CCIS 81, pp. 30–43, 2010, ISBN 3-642-14057-2 Springer Berlin Heidelberg NewYork, ISSN 1865–0929.

Schlimmer, J.C. and Fisher, D.H. (1986) A case study of incremental concept induction. *In AAAI*, pp. 496–501.

Stenger, B.V. *et al.* (2001) *Topology-free Hidden Markov Models: Application to Background Modelling*. Proc. of IEEE Conference on Computer Vision.

Stilwell, D.J. and Bishop, B.E. (2000) Platoons of underwater vehicles. *IEEE Control Systems Magazines*, **20** (6), 45–52.

Sugeno, M. and Kang, G.T. (1988) Structure identification of fuzzy model. *Fuzzy Sets and Systems*, **28** (1), 15–33.

Sutton, R.S. and Barto, A.G. (1999) Reinforcement learning. *Journal of Cognitive Neuroscience, MIT*, **11** (1), 126–134.

Takagi, T. and Sugeno, M. (1985) Fuzzy identification of systems and its application to modeling and control. *IEEE Transactions on Systems, Man & Cybernetics*, **15** (1), 116–132.

Trevisan, J. *et al.* (2010) Syrian hamster embryo (SHE) assay (pH 6.7) coupled with infrared spectroscopy and chemometrics towards toxicological assessment. *Analyst*, **135** (12), 3266–3272.

Trevisan, J. *et al.* (2012a) Extracting biological information with computational analysis of Fourier-transform infrared (FTIR) biospectroscopy datasets: current practices to future perspectives. *Analyst*, **137** (14), 3202–3215.

Trevisan, J. *et al.* (2012b) IRootLab: an open-source MATLAB Toolbox for biospectroscopy data analysis, *Bioinformatics*, to appear.

Tsymbal, A. (2004) The problem of concept drift: definitions and related work. *Technical Report TCD-CS-2004-15*, Department of Computer Science, Trinity College Dublin, Ireland.

Tsypkin, Y. (1968) Self-learning–What is it?, *IEEE Transactions on Automatic Control*, **13** (6), 608–612.

Valavanis, K. (2006) *Unmanned Vehicle Navigation and Control: A Fuzzy Logic Perspective, Proc. of the 2006 International Symposium on Evolving Fuzzy* Systems, Ambleside, Lake District, UK, IEEE Press, pp. 200–207.

Vapnik, V.N. (1998) *The Statistical Learning Theory*, Springer Verlag, Berlin, Germany.

Wang, L.-X. (1992) *Fuzzy Systems are Universal Approximators.* Proc. of the First IEEE International Conference on Fuzzy Systems, FUZZ-IEEE – 1992, San Diego, CA, USA, IEEE Press, pp. 1163–1170.

Wang, L.X. (1994) *Adaptive Fuzzy Systems and Control*, Prentice Hall Inc, USA.

Werbos, P.J. (1974) *Beyond Regression: New Tools for Prediction and Analysis in the Behavioral Sciences*, PhD thesis, Harvard University.

Werbos, P.J. (1990) Back-propagation through time: what it does and how to do it. *Proceedings of the IEEE*, **78** (10), 1550–1560.

Widmer, G. and Kubat, M. (1996) Learning in the presence of concept drift and hidden contexts. *Machine Learning*, **23** (1), 69–101.

Widrow, B. and Stearns, S. (1985) *Adaptive Signal Processing*, Prentice Hall, Englewood Cliffs, NJ, USA.

Woodbury, M.A. (1950) *Inverting Modified Matrices*, Memorandum Rept. 42, Statistical Research Group, Princeton University, Princeton, NJ, p. 4.

Yager, R. (2006) *Learning Methods for Intelligent Evolving Systems*. In Proc. 2006 International Symposium on Evolving Fuzzy Systems, Ambleside, UK, IEEE Press, pp.3–7.

Yager, R.R. (1988) On ordered weighted averaging aggregation operators in multi-criteria decision making, *IEEE Transactions on Systems, Man and Cybernetics*, **18** (1), 183–190.

Yager, R.R. and Filev, D.P. (1993) *Learning of Fuzzy Rules by Mountain Clustering*. Proc. of the SPIE Conf. on Application of Fuzzy Logic Technology, Boston, MA, USA, pp. 246–254.

Yager, R.R. and Filev, D.P. (1994) *Essentials of Fuzzy Modeling and Control*, John Wiley and Sons, New York, USA.

Yen, J. and Gillespie, W. (2002) *A Global Local Learning Algorithm for Identifying Takagi-Sugeno-Kang Fuzzy Models*. Proc of VIEEE International Conference on Fuzzy Systems, pp. 967–972.

Yuan, Y. and Shaw, M.J. (1995) Induction of fuzzy decision trees. *Fuzzy Sets and Systems*, **69**, 125–139.

Zadeh, L.A. (1965) Fuzzy sets. *Information and Control*, **8** (3), 338–353.

Zadeh, L.A. (1975) The concept of a linguistic variable and its application to approximate reasoning – I. *Information Sciences*, **8** (3), 199–249.

Zhivkovic, Z. and Van der Heijden, F. (2006) Efficient adaptive density estimation per image pixel for the task of background subtraction. *Pattern Recognition Letters*, **27** (7), 773–780.

Zhou, X. and Angelov, P. (2007) *An Approach to Autonomous Self-Localization of a Mobile Robot in Completely Unknown Environment Using Evolving Fuzzy Rule-Based Classifier*. Proc. First 2007 IEEE Intern. Symposium on Computational Intelligence Applications for Defense and Security, IEEE Symposium Series on Computational Intelligence, SSCI-2007, Honolulu, Hawaii, USA, pp. 131–138.

Zhou, X.-W. and Angelov, P. (2006) *Real-Time joint Landmark Recognition and Classifier Generation by an Evolving Fuzzy System*. Proc. of the 2006 IEEE World Congress on Computational Intelligence, WCCI-2006, Vancouver, BC, Canada, pp. 6314–6321.

Glossary

AI	artificial intelligence
AIS	autonomous input selection
AL	autonomous learning
ALS	autonomous learning systems
ACS	automatic control system
ANFIS	adaptive neuro-fuzzy inference systems
AnYa	one of the three main types of FRB
ARCOS	advanced robotics control and operations firmware
ARIA	advanced robot interface for applications (software package in C)
ART	adaptive resonance theory
AutoClassify	autonomous classifier
AutoCluster	autonomous clustering
AutoControl	autonomous controller
AutoPredict	autonomous predictor
AS	autonomous system
ASTM	American Society for Testing and Materials
AutoSense	autonomous sensor
BS	background subtraction
CCTV	closed-circuit television
CPU	computer processing unit
DARPA	Defense Advanced Research Projects Agency
DENFIS	dynamically evolving fuzzy inference systems
DM	decision making
DSS	decision support systems
EA	evolutionary algorithm
EBP	error backpropagation method (also known as the delta rule)
EFM	evolving fuzzy models

Autonomous Learning Systems: From Data Streams to Knowledge in Real-time, First Edition. Plamen Angelov.

EFS	evolving fuzzy systems
EKF	extended Kalman Filter
ELM	evolving local means
eIS	evolving intelligent system
eTS	evolving Takagi–Sugeno type fuzzy rule-based system
EO	electro-optical
FCM	fuzzy C means
FL	fuzzy logic
FLC	fuzzy logic controller
FN	false negative
FP	false positive
FRB	fuzzy rule-based system
FTIR	Fourier transform infrared (spectroscopy)
GA	genetic algorithm
GFS	genetic fuzzy systems
GMM	Gaussian mixture models
GK	Gustafson–Kessel clustering method
GP	genetic programming
GPS	global positioning system
GSM	Groupe Speciale Mobile, also Global System for Mobile Communications
HGO	heavy gas oil
HN	heavy naphtha
HUV	hue, saturation and value (a representation in image processing)
HMM	hidden Markov models
ICA	independent component analysis
iLVQ	incremental learning vector quantiser
IR	infrared
KDE	kernel density estimation
kNN	k nearest neighbour
KNO	kerosene oil
LDA	linear discriminant analysis
LGO	light gas oil
LMS	least mean squares
LS	least squares
LVQ	learning vector quantiser
MGO	medium gas oil
MIMO	multi-input–multi-output system
MINO	multiple-input, no output
MISO	multi-input–single-output
ML	machine learning
MoM	mean of maxima
MoD	Ministry of Defence, UK
NDEI	nondimensional error index

NFS	neuro-fuzzy system
NN	neural network
P	proportional controller
PI	proportional integral controller
PC	principle component
PCA	principle component analysis
PDF	probability density function
PLC	partial least squares
PR	pattern recognition
RBF	radial basis function
RDE	recursive density estimation
RES	atmospheric residue
RLS	recursive least squares
RMSE	root mean square error
rPCA	recursive PCA
rLDA	recursive LDA
SA	statistical analysis
SAR	singular aperture radar
SLAM	self-localisation and mapping
SI	system identification
SPARC	self-organising parameter-free autonomous rule-based controller
SOM	self-organising map
SVM	support vector machines
SVD	singular value decomposition
TS	Takagi–Sugeno-type fuzzy rule-based systems
UAS	uninhabited/unmanned aerial/airborne system
UAV	unmanned aerial vehicle
UGV	unmanned ground-based vehicle (mobile robot)
UxV	unmanned vehicles
VQ	vector quantisation
wRLS	(fuzzily) weighted RLS

Index

Autonomous Learning Systems: From Data Streams to Knowledge in Real-time, First Edition. Plamen Angelov.